"Philosophy is written in that great book which ever lies before our eyes, I mean the universe, but we cannot understand it if we do not first learn the language and grasp the symbols in which it is written. This book is written in the mathematical language..."

Galileo Galilei.

What is the next number in this sequence:

7, 22, 11, 34, 17, 52, 26, 13, 40, 20, 10, 5, 16, ...

Other titles from the Simplicity Research Institute

Simplicity in Complexity
by Rajesh R. Parwani

Real World Mathematics
by Wei Khim Ng and Rajesh R. Parwani

Available from the online SRI Bookstore
www.store.simplicitysg.net
and other outlets

Integrated Mathematics

for

Explorers

Intermediate

Adeline Ng
Cherish Academy, Singapore

Rajesh R. Parwani
SRI, Singapore

Simplicity Research Institute, Singapore
www.simplicitysg.net

Integrated Mathematics for Explorers.
Published by the Simplicity Research Institute, Singapore.

Copyright © 2013, by Adeline Ng and Rajesh R. Parwani.

All rights reserved. No part of this book may be reproduced in any form, stored in any information retrieval system, or transmitted by any means, except as permitted by provisions of the Copyright Act, without written permission of the publisher.

For bulk orders, special discounts, or to obtain customised versions of this book, please contact the Simplicity Research Institute at **enquiry@simplicitysg.net**.

A CIP record for this book is available from the National Library Board, Singapore.

Print Edition SRI-2013-3A
(Errata Corrected Dec 2015)
ISBN: 978 − 981 − 07 − 8234 − 4

Contents

Preface

0.1 Overview

We have written this book for mathematics lovers: A compact resource covering a range of topics, mostly at an intermediate level[1].

Each topic in this book is introduced with a brief motivation or application, with several real-life situations illustrated in the practice questions.

The practice questions have been divided into two categories: The **Exercises** are meant to develop basic understanding and build confidence. At the next level we have **Problems**, which typically require greater analytical skills or examine multiple concepts. Some of the problems are much more complex than others but we have decided not to distinguish them by additional labels (sometimes, what you do not know might be good for you).

To assist students, we have provided numerous **worked examples**, which however still require the student to participate actively in filling some minor gaps.

For readers who are keen to venture beyond the boundaries of a typical school syllabus, we have included a number of questions in the **Challenges** category.

---

[1] Typically, for those engaged in the middle stage of an integrated programme, the International Baccalaureate, or the IGCSE. However, we hope others too will enjoy the material here.

The **Inquire and Investigate** section in each chapter is for readers who would like to engage in project work or self-study; we hope it may serve as a springboard for their endeavours.

Puzzle solving, or recreational mathematics, is an enjoyable and stimulating activity for most students. We have included a number of classic **puzzles** in a separate chapter at the end of the book. Though most of the puzzles do not require much technical knowledge, they do encourage lateral experimentation and logical thinking, and more importantly, they are fun!

Sometimes puzzles remain unsolved for a long time despite the attention they have received from amateur and professional mathematicians. In the Escapades chapter we introduce the interested reader to the **research frontier** by using some open problems which are easy to describe.

Mathematics is a human activity, and to remind the reader of this we have included mini-biographies and quotations of some prominent mathematicians and scientists.

We intend to provide a number of e-resources, such as supplementary chapters on topics not included in this hard copy, on the book's website:

www.simplicitysg.net/books/imaths

We hope our approach will encourage more readers towards enquiry and exploration, and help them integrate the knowledge they have gained.

If you are not yet a mathematics lover, we hope you will be seduced by something in this book.

0.2 Guide for Students

Explore....Often, there is more than one way to solve a mathematics problem.

Test your understanding of new concepts, and develop your mathematical skills, by trying the relevant exercises, using the worked examples as reference where necessary. We have left gaps in many of the worked examples for you to fill as you review them.

Answers to all the exercises are provided at the end of the book. We have also included hints to some of the exercises in an appendix.

Once you have a good understanding of the basic material in a chapter, try your hand at some of the questions in the Problems category. Answers to the odd-numbered problems are at the end of the book.

By omitting answers to the even-numbered problems we hope to encourage peer discussion, which is generally beneficial. We plan to provide hints or answers to the even-numbered problems on the book's website www.imaths.info.

We have included a large number of practice questions under each topic to provide you with variety and extensive practise. However, you probably do not need to solve all the questions to master the subject. (For those who desire even more questions, you can find them in the Supplementary chapter.)

We have provided proofs to support some of the theorems. However, on first reading you may skip these and focus on the application of the theorems in the practice questions.

The Challenges and Inquire and Investigate sections of each chapter provide you with opportunities for further study and exploration, either on your own or with the help of a mentor.

Need a diversion or inspiration? Browse through the Puzzles and Research sections in the Escapades chapter.

Finally, throughout this book we have included some tips on relaxing, de-stressing and re-energising yourself; you will find them when you need them.

0.3   To Teachers

We have organised the material in this book in a consolidated manner. For example, merging topics like "indices" and "exponents", and treating quadratic equations and higher-degree polynomials in one chapter.

You can choose to present the material in a different order and correspondingly guide the student as to the questions they should try in each chapter. There are some questions which require knowledge from different chapters; instead of organising things very strictly in a linear manner, this book gives you the flexibility to experiment and discover what suits your needs.

We hope the Challenges and Inquire and Investigate sections of each chapter will provide you with more ideas to inspire the adventurous students in your classes.

We intend to provide you with supplemental support over time through various channels including workshops and e-supplements.

Please check our website, www.simplicitysg.net/books/imaths, for updates and if you have any queries, please send them to **enquiry@simplicitysg.net**.

0.4   Notation and Convention

We sometimes refer to specific worked examples or practice questions in the main text. The prefixes, E, P, C, I, WE and Q refer respectively to Exercise, Problem, Challenge, Inquire and Investigate, Worked Example, and Question.

The summation notation is often useful:

$$\sum_{r=1}^{r=N} a_r \equiv a_0 + a_1 + ... + a_r + ... + a_N \ ,$$

as is the "overdot" for the time-derivative: $\dot{x} \equiv \dfrac{dx}{dt}$ and $\ddot{x}$ for the second time derivative.

Unless otherwise stated, all constants and variables in this text refer to real numbers.

Often we have not specified the units for lengths, for example, writing $a = 5$ for the side of triangle; the reader should read it as 5 "units".

0.5   Acknowledgements

The quotations in this book have been extracted from various sources, such as Wikipedia and The MacTutor History of Mathematics website. Similarly, the mini-biographies in this book were compiled using information from the above two sites and various books the authors have read over the years.

We are grateful to all those who read parts of our draft manuscript and provided their valuable feedback. Among them are: Max Shivan, Koh King Koon, Li Lingzhi and Parizad Setna.

Adeline thanks Dennis Ng, Cherish Finden Ng, Ng Yoke Fung and Tan Cheng Kiong for their support, encouragement and inspiration during the various stages in the making of this book.

Rajesh thanks his father, Ramchand N. Parwani, for feeding him stimulating mathematics problems during his formative years; some of them have been reproduced in this book.

Adeline Ng

Rajesh Parwani

October 2013, Singapore.

What did one mathematics book say to the other?

Don't bother me, I have my own problems!
(Anonymous, from the Web)

Chapter 1

Exponents and Logarithms

Now I will have less distraction.
Leonhard Euler

1.1 Introduction

Consider the repeated multiplication $7 \times 7 \times 7$. A shorthand for that product is 7^3, a notation that is not only compact but also allows for rapid calculations using the rules that we summarise below. More generally, if one writes $y = b^x$ for $b \neq 1$ and $b > 0$, then b is called the **base** while x is the **exponent** which need not be an integer. The function $y = b^x$ is sometimes referred to as an indicial function and the exponent x as the index or power.

Integral exponents occur in our familiar decimal notation which uses the base 10, for example 352 is $3 \times 10^2 + 5 \times 10^1 + 2 \times 10^0$. That same number may be written in **scientific notation** as 3.52×10^2. (More generally, any nonzero number may be written in the form $\pm N.a_1 a_2 a_3.... \times 10^p$ where $1 \leq N \leq 9$ is an integer, p an integer and the a_i (for $i = 1, 2...$) are non-negative integers.)

Fractional exponents are useful too. For example $2^{1/2} = \sqrt{2}$. Numbers such as $\sqrt{N}$, where N is not a perfect square of an integer, are sometimes called **surds**.

Given the function $y = b^x$, the inverse relation is written $x = \log_b y$ where 'log' is short for the **logarithm** function, in this case to base b.

That is,

$$y = b^x \Leftrightarrow x = \log_b y \ . \tag{1.1}$$

The choice $b = 10$, the **common logarithm**, was the first to be studied by Napier while the **natural logarithm**, introduced by Euler, has the irrational base $e = 2.718281....$ The notation $\log_e$ is often abbreviated as ln, while $\log_{10}$ is sometimes written as lg.

Although the choice of the irrational number e as a base for logarithms might seem odd, such a choice actually leads to function with nicer mathematical properties (see the Calculus chapter later). Furthermore, the exponential function e^x occurs in many mathematical laws describing natural phenomena as we will see in the practice questions.

Historically, logarithms were introduced to simplify multiplication but they have useful properties which make them interesting in their own right. One practical application of logarithms is in compressing a large range of data values; for example, if a variable x takes values from 10^{-10} to 10^{10} then $\log_{10} x$ ranges only from -10 to 10.

The Richter scale describing earthquakes is a logarithmic scale with base 10, meaning that a 5.0 earthquake releases 10 times more energy than a 4.0 earthquake – can you work out why this is the case?

Leonhard Euler

The leading mathematician of the 18th century, Euler was both creative and prolific, making important contributions to many areas of mathematics and theoretical physics. Even when he was almost blind in the last twenty of his life, he continued his research unabated, relying on his amazing memory and power of mental calculation. His name is associated with many ideas, such as the bridges of Konigsberg problem (see the Escapades chapter), and the remarkable Euler identity $e^{i\pi} + 1 = 0$, where $i = \sqrt{-1}$ (see the Escapades chapter).

1.2   Relations and Properties

We list here some useful identities involving exponents. We will not prove them but encourage you to check their plausibility using integer exponents, see also the practice questions (P6, C1).

For $a, b > 0$,

$$a^{-1} = \frac{1}{a}. \tag{1.2}$$

$$a^{xy} = (a^x)^y. \tag{1.3}$$

$$a^x a^y = a^{x+y}. \tag{1.4}$$

$$(ab)^x = a^x b^x. \tag{1.5}$$

It is useful to note that for $a > 1$, $y = a^x$ is positive and an increasing function of x along the real line. This implies that for $0 < a < 1$, a^x is a decreasing function of x (can you show this?).

From the rules for exponents you can easily deduce the basic operation involving surds: $\sqrt{ab} = \sqrt{a}\,\sqrt{b}$.

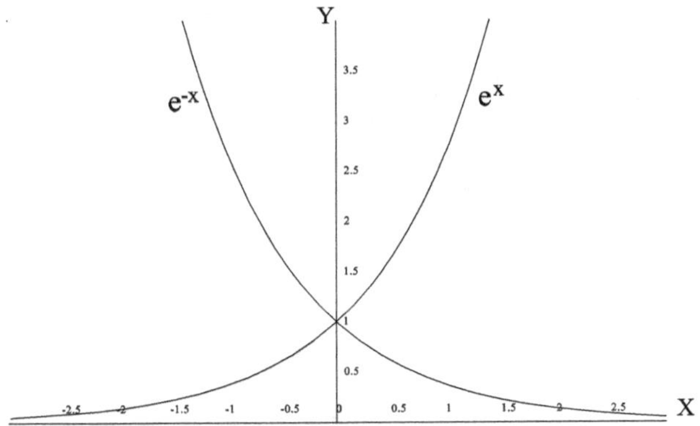

Figure 1.1: The exponential functions e^{-x} and e^x cross at $x = 0$, $y = 1$.

Since the logarithm is the inverse function to the indicial function, see (1.1), the relevant identities may be derived from those for exponents above, see also practice questions (C1, C2). For $a, b > 0, a \neq 1, b \neq 1$ and $P, Q > 0$,

$$\log_b PQ = \log_b P + \log_b Q. \tag{1.6}$$

$$\log_b \frac{P}{Q} = \log_b P - \log_b Q. \tag{1.7}$$

$$\log_b P^c = c \log_b P. \tag{1.8}$$

$$\log_b P \;=\; \frac{\log_a P}{\log_a b}\,. \tag{1.9}$$

$$\log_b a \;=\; \frac{1}{\log_a b}\,. \tag{1.10}$$

For $a > 1$ and $x > 0$, $y = \log_a x$ is an increasing function of x; it is positive for $x > 1$, negative for $x < 1$ and vanishes at $x = 1$. Can you work out the behaviour for $0 < a < 1$?

Note that since $y = \log_a x$ is the inverse function to the function $y = a^x$, its graph may be obtained by reflecting the exponential curve about the line $y = x$ (why?).

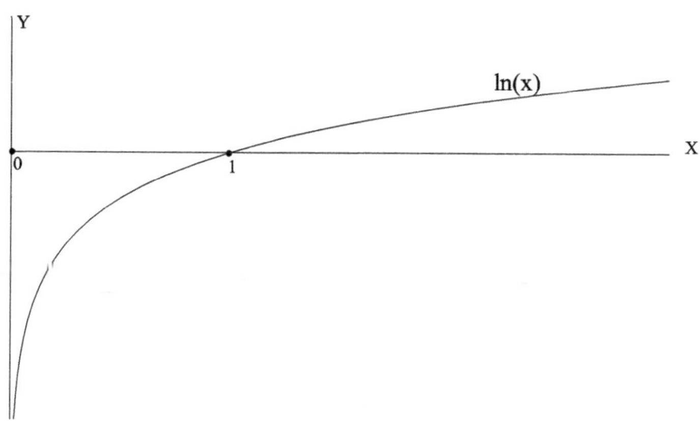

Figure 1.2: The natural logarithm.

1.2.1   Binomial Expansions

We have seen that $(ab)^n = a^n b^n$. Is there also a different way of expressing $(a + b)^n$? Let us look at some special cases first. We know $(a + b)^2$ is $a^2 + 2ab + b^2$ and similarly we can work out, by direct multiplication, $(a + b)^3 = a^3 + 3a^2 b + 3ab^2 + b^3$. Do you notice a pattern emerging?

More generally, for a positive integer n, we have the **Binomial Theorem**,

$$(a+b)^n = a^n + \binom{n}{1}a^{n-1}b + \binom{n}{2}a^{n-2}b^2 + \binom{n}{3}a^{n-3}b^3 + \ldots$$
$$+ \binom{n}{r}a^{n-r}b^r + \ldots + \binom{n}{n-1}ab^{n-1} + b^n \tag{1.11}$$

where the binomial coefficient

$$^nC_r \equiv \binom{n}{r} = \frac{n!}{r!(n-r)!} \tag{1.12}$$

counts the number of ways of choosing r objects from n identical ones. The factorial symbol is defined by $n! = n \times (n-1) \times (n-2) \times \ldots \times 2 \times 1$ with $0! \equiv 1$.

Note that the $(r+1)$-th term in the above expansion is given by

$$\binom{n}{r}a^{n-r}b^r. \tag{1.13}$$

The binomial theorem is easy to verify once the left hand side is written as n products. That is, $(a+b)^n = (a+b)(a+b) \ldots (a+b)$. Then clearly in the expansion there is only one way to form the a^n term.

Next, the $a^{n-1}b$ term is formed when b is chosen from exactly one bracket and there are $\binom{n}{1}$ ways to do this. Likewise the $a^{n-r}b^r$ term can be formed in $\binom{n}{r}$ ways.

The binomial theorem is used in many contexts, such as for making approximate calculations, and in probability theory.

1.3 Worked Examples

1. *Evaluate $P = (1.02 \times 10^2) \times (3.6)^3 \times e^4$ by three methods: (a) Directly using a calculator, (b) first evaluating $\log_{10} P$, (c) first evaluating $\ln P$.*

Solution:

(a) $P = 2.598 \times 10^5$ (correct to four significant figures).

(b) We have

$$
\begin{aligned}
\log_{10} P &= \lg(1.02 \times 10^2 \times 3.6^3 \times e^4) \\
&= \lg 1.02 + \lg 10^2 + \lg 3.6^3 + \lg e^4 \\
&= \lg(1.02) + 2 + 3\lg(3.6) + 4\lg(e) \\
&= 5.4147
\end{aligned}
$$

which implies $P = 10^{5.4147} = 2.598 \times 10^5$.

(c) Similarly, $\ln P = 12.468$ and so $P = e^{12.468} = 2.598 \times 10^5$.

Answer: $P = 2.598 \times 10^5$ (correct to four significant figures).

---

2. *Re-arrange the following sequence in order from the smallest to the largest:* $\log_2 3,\ \log_3 2,\ \log_2 6$.

Solution:

In order to compare the magnitudes we need to convert all of them to the same base. Using eq.(1.10), the middle value is $\log_3 2 = 1/\log_2 3$. Now, since $\log_2 x$ is an increasing function of x and $\log_2 2 = 1$, therefore $\log_2 3 > 1$ and $\log_3 2 = 1/\log_2 3 < 1$.

Similarly, $\log_2 6 > \log_2 3$. More precisely,
$\log_2 6 = \log_2 2 \cdot 3 = \log_2 2 + \log_2 3 = 1 + \log_2 3$.

Therefore we have the ordering $\log_3 2 < 1 < \log_2 3 < \log_2 6$.

Answer: $\log_3 2,\ \log_2 3,\ \log_2 6$.

---

3. *Find x in terms of $a > 0$ and b if $\log_a(x^2 - 2b) - \log_{\sqrt{a}}(x-1) = 0$.*

Solution:

This might look complicated, but let us focus first on converting all the logarithm's to the same base. Using eq.(1.9), for any Y we have $\log_{\sqrt{a}} Y = \dfrac{\log_a Y}{\log_a \sqrt{a}} = 2\log_a Y$ since $\log_a \sqrt{a} = \log_a a^{1/2} = 1/2$.

Therefore the equation we need to solve is
$\log_a(x^2 - 2b) = 2\log_a(x - 1) \Rightarrow \log_a(x^2 - 2b) = \log_a(x - 1)^2$.
This implies $(x^2 - 2b) = (x-1)^2$ which easily yields $x = (1+2b)/2$.
Notice that there is no dependence on a in the final result.

Answer: $x = (1 + 2b)/2$.

---

4. *Solve for x in the equation $49^x + 7^{2x-1} + 7^x = 7^{-1}$ giving your answer to two decimal places.*

Solution:

The unknown x appears as an exponent of various bases. So we first try to identify a common base. Since $49 = 7^2$, we are led to re-write the equation as $7^{2x} + 7^{2x} \cdot 7^{-1} + 7^x = 7^{-1}$ which suggests the natural substitution $y = 7^x$ giving $8y^2 + 7y - 1 = 0$; the solutions of this quadratic equation are $y = -1$ and $y = 1/8$.

However, since $y = 7^x > 0$ for real x, only $y = 1/8$ is acceptable. So $7^x = 1/8$, and taking logarithms of both sides gives the solution $x = -\log 8/\log 7 = -1.07$.
(Which base should you choose for the logarithm in the last equation? It does not matter! You can choose base 10 or e, or whatever is convenient for you. See eq. (1.9) which shows that the result does not depend on the base.)

Answer: $x = -1.07$.

---

5. *The curve described by $ay - bx^n - 3 = 0$ is known to pass through the points $(0,3)$, $(1,5)$ and $(2,35)$. Determine the constants a, b, n.*

Solution:
This problem involves three unknowns and the given points give three conditions. In general, such a problem would involve three simultaneous equations if all the unknowns were coupled in the three conditions, but in this particular case the first two points easily give us a and b : First, $3a - 0 - 3 = 0 \Rightarrow a = 1$, then $5 \times 1 - b - 3 = 0 \Rightarrow b = 2$.

The final point gives $35 - 2 \times 2^n - 3 = 0 \Rightarrow 2^n = 2^4$, so $n = 4$.

Answer: $a = 1$, $b = 2$, $n = 4$.

---

6. Simplify $\dfrac{\sqrt{2}}{2\sqrt{2} - \sqrt{3}}$.

Solution:
The standard procedure to rationalise the denominator, that is to remove the surds, is to use the identity $(a - b)(a + b) = a^2 - b^2$. In this case we multiply both the numerator and denominator by $2\sqrt{2} + \sqrt{3}$ and simplify the terms.

Answer: $\dfrac{4 + \sqrt{6}}{5}$.

---

7. *Solve for x in the equation $1 + \sqrt{x - 4} = \sqrt{12 - x}$ expressing your answer correct to two decimal places.*

Solution:
Squaring both sides and collecting terms gives $2\sqrt{x - 4} = 15 - 2x$.

We square this again to get $x^2 - 16x + 60.25 = 0$, the solutions of which are $x = 9.94$ and $x = 6.06$.

However, because we have squared the original equation (twice in fact!), we should check whether the two values of x we have obtained also satisfy the original equation. Substitution shows that while $x = 6.06$ is a valid solution, $x = 9.94$ is not!

(In order to understand how and why a **spurious solution** can arise, it is useful to look at a simpler example: $2y = y + 1$; obviously the unique solution to this linear equation is $y = 1$. But if you square both sides (try it) you obtain a quadratic equation which gives not only the solution $y = 1$, but also the spurious solution $y = -1/3$ which does not satisfy the original linear equation.

Now returning to our original problem, you can check that if the signs in front of the square roots were changed, so that the equation was $1 - \sqrt{x-4} = -\sqrt{12-x}$, the squaring procedures would have yielded the same two solutions $x = 9.94$ and $x = 6.06$ but now only $x = 9.94$ satisfies the starting equation.)

Answer: $x = 6.06$.

---

8. *Find (a) the coefficient independent of x, and (b) the coefficient of the x^3 term, in the expansion of $y = x(1 + 3x)\left(2x + \frac{1}{x}\right)^8$.*

Solution:
Write the expression as $y = (x + 3x^2)\left(2x + \frac{1}{x}\right)^8$. The second bracket has the binomial expansion

$$\left(2x + \frac{1}{x}\right)^8 = \sum_{r=0}^{r=8} \binom{8}{r}(2x)^{8-r} x^{-r} \tag{1.14}$$

$$= \sum_{r=0}^{r=8} \binom{8}{r}(2)^{8-r} x^{8-2r} , \tag{1.15}$$

$$\equiv \sum_{r=0}^{r=8} T_{r+1} \qquad (1.16)$$

which involves only *even* powers of x; the powers become negative for $r > 4$.

In other words, the $(r+1)$-th term in the expansion of the second bracket in y is $T_{r+1} = \binom{8}{r}(2)^{8-r}x^{8-2r}$

(a) The coefficient independent of x in y will be the term x^0 in the expansion, which can only arise when $3x^2$ multiplies a term from the second bracket. Consider therefore $3x^2 \cdot T_{r+1} = 3\binom{8}{r}(2)^{8-r}x^{10-2r}$ and set $10 - 2r = 0 \Rightarrow r = 5$. Thus the term independent of x in y is $3 \cdot \binom{8}{5}2^{8-5} = 1344$.

(b) Similarly, to get the coefficient of the x^3 term in y we have to look at $xT_{r+1} = \binom{8}{r}(2)^{8-r}x^{9-2r}$ and choose $r = 3$. The coefficient is thus $\binom{8}{3}2^{8-3} = 1792$.

Answers: (a) 1344, (b) 1792.

1.4   Exercises

1. Express the following numbers in scientific notation:

   (a) -0.0031.

   (b) 123.456.

   (c) $3.23 \times 12.56 \times 0.02$.

   (d) $-24.31 \times 10^5 \times 0.07 \times 10^{-6}$.

2. The distances of the Moon, Sun and the star Proxima Centauri from Earth are respectively 3.8×10^5, 1.5×10^8 and 4.0×10^{13} kilometres (km). Taking the speed of light to be 3.0×10^5 km/s,

calculate the time it takes light to reach Earth from each of those bodies. Express your answer in seconds for the Moon, in minutes for the Sun and in years for Proxima Centauri. (Note: speed = distance travelled / time taken.)

3. A convenient way to compare the intensity I_1 of a sound to a reference intensity I_0 is through the decibel (dB) value defined by $L = 10 \log_{10} \frac{I_1}{I_0}$.
   (a) If $L = 20$ dB, what is the ratio of the two sound intensities?
   (b) If $I_1 = 2I_0$, what is the dB value?
   (c) Optional: What is the maximum dB value beyond which sustained exposure may harm the human ear?

4. When radioactive atoms decay, they change into atoms of a different type. If the initial number of radioactive atoms is N_0, then the number that remain unchanged at time t is given by the equation $N(t) = N_0 e^{-\lambda t}$ where λ, called the "decay constant", depends on the element.
   (a) Determine, in terms of λ, the time taken for the number of radioactive atoms of a specific type to drop to half its starting value. (This value, denoted by $t_{1/2}$, is called the half-life of that element.)
   (b) A scientist estimates that the amount of (radioactive) Carbon-14 remaining in a sample is 1/4 of its original value. Given that the half-life of Carbon-14 is 5730 years, estimate the age of the sample.
   (c) The scientist finds another radioisotope in a different sample. He estimates the second sample to be 75,000 years old and the radioisotope present to be at 12% of its original value. What is the half-life of the radioisotope?
   (d) Optional: What could the second radioisotope be?

5. In Einstein's Special Theory of Relativity, the energy of a particle moving at speed v is given by $E = \dfrac{mc^2}{\sqrt{1 - v^2/c^2}}$ where c is the speed of light and m the mass of the particle.
   (a) What would the energy of the particle be when $v = 0$?
   (b) If E and m are real and finite, what does Einstein's expression

imply for the maximum speed of a particle when $m > 0$?
(c) Optional: If E is finite, show that one may have $v = c$ only if $m = 0$. Do such particles exist?

6. Rydberg's formula for hydrogen describes the wavelength, λ, of radiation emitted in transitions of the atom. With integers $n_2 > n_1 \geq 1$, the formula is $\dfrac{1}{\lambda} = R\left(\dfrac{1}{n_1^2} - \dfrac{1}{n_2^2}\right)$ where $R = 1.097 \times 10^7 \ m^{-1}$ is Rydberg's constant.
(a) Calculate the longest wavelength that a hydrogen atom can emit in the $n_1 = 1$ (Lyman) series.
(b) Calculate the shortest wavelength that a hydrogen atom can emit in the $n_1 = 2$ (Balmer) series.
(c) Optional: Are any of the emissions in parts (a) and (b) visible to the human eye?

7. Re-arrange each of the following sequences in order, from the smallest to the largest, without using a calculator. Justify your answers.
(a) 7^5, $5^{7/2}$, 7^7, 5^5.
(b) $\log_2 3$, $\log_2 \sqrt{3}$, $\log_2 9$, $\log_2 \dfrac{1}{3}$,

8. Re-arrange each of the following sequences in order, from the smallest to the largest, without using a calculator. Justify your answers.
(a) $8^{1/3}$, $4^{-\frac{3}{2}}$, $\left(\dfrac{1}{2}\right)^{-2}$.
(b) $\left(2^6\right)^{\frac{1}{3}}$, $\left(2^{\frac{1}{2}}\right)^{-4}$, $2^2 \times 2^{-3}$, $\left(2^2 \times 3^4\right)^{-\frac{1}{2}}$.

9. Simplify each of the following expressions as much as possible without using a calculator:
(a) $\left(2x^2\right)^{-\frac{1}{2}} (3x^{-5})^{-1}$.
(b) $\log_5 7 + 7\log_5 \dfrac{1}{49} + \log_5 \sqrt{35}$.
(c) $\log_3 25 + \dfrac{1}{\log_5 27}$.

10. Write each of the following expressions in the form $a + b\sqrt{c}$ with a, b rational numbers and c an integer which is as small as possible:

(a) $2\sqrt{27} + \sqrt{\dfrac{3}{4}}.$ (c) $5\sqrt{8} - \dfrac{1}{\sqrt{72}} + \sqrt{50}\,\sqrt{\dfrac{9}{2}}.$

(b) $\dfrac{5 - 2\sqrt{7}}{\sqrt{7} + 3}.$

11. Solve each of the following equations for x (where possible, avoid using a calculator):

   (a) $5^{2x-1} = 1/25.$

   (b) $3^{2-x}(2^2 \times 3^{2x+1}) = 4/9.$

   (c) $\log_3(2x - 1) = 2.$

   (d) $\log_{10}(x + 1) - \log_{10}(x - 1) = 2 - \log_{10} 2.$

   (e) $\dfrac{\log_2 x^2}{2 \log_x 2} = 9.$

   (f) $\sqrt{x - 1} = x - 3.$

   (g) $7^3 = e^x.$

12. Use the Binomial Theorem to estimate each of the following to one decimal place, without using a calculator:

   (a) $(1.02)^4.$ (b) $(2.02)^6.$

13. Find the first three terms, in increasing powers of x, in the expansion of each of the following expressions:

   (a) $\left(1 + \dfrac{x}{2}\right)^6.$ (d) $(1 + x)^2(2 + x)^6.$

   (b) $(1 + 2x^2)^9.$ (e) $x(1 + 3x)\left(1 + \dfrac{1}{x}\right)^6.$

   (c) $\left(x + \dfrac{1}{x}\right)^8.$

14. (a) Sketch the functions $f(x) = 3\ln x$ and $g(x) = 1 - 2x$ on the same graph to determine how many solutions there are to the

equation $3\ln x + 2x - 1 = 0$.
(b) By using an accurate plot, estimate those solutions.

15. (a) Sketch the functions $f(x) = 2^x$ and $g(x) = 8x(1-x)$ on the same graph and use it to determine how many solutions there are to the equation $x - x^2 - 2^{x-3} = 0$ in the range $0 \leq x \leq 1$.
(b) By using an accurate plot, estimate the solution that lies in the range $0 \leq x \leq 1/2$.

1.5   Problems

1. In the field of microelectronics there is an empirical observation known as Moore's Law. It states that the number of transistors on integrated circuits doubles every two years.
(a) Express the law in mathematical form and use it to estimate the time needed for the number of transistors on integrated circuits to increase ten-fold since the start.
(b) Optional: Is the law still accurate nowadays?

2. According to one legend, a rich king decided to reward his court mathematician for inventing the game of chess which the king enjoyed playing. He asked the mathematician to state his desired reward. The mathematician replied: "I would like some grains of rice for each square of the chessboard, one for the first square, two for the second, four for the third, and so on, with each square doubling the grains of rice compared to the previous square." The king was amused by the seemingly modest request and agreed; he instructed his treasurer to make the necessary payment. Do you think the treasurer would have been able to make the payment? (Note: A chessboard has 64 squares.)

3. A microbiologist observes that the bacterial cells in his dish double in number every 12 minutes, starting with 5 cells at $t = 0$.
(a) How many cells are there in the dish at the end of one hour?
(b) How long does it take the bacterial colony to reach 2000 cells?

4. The curve formed by a freely hanging cable supported only at its ends is called a catenary. Its equation is $y = \dfrac{a}{2}\left(e^{x/a} + e^{-x/a}\right)$.

(a) Sketch the curve.

(b) If $a = 5$ m and the supports of the cable are at $x = \pm 20$ m, determine how low the cable reaches below the support level.

5. Re-arrange each of the following sequences in order, from the smallest to the largest, without using a calculator. Justify your answers.

(a) $\log_2 3, \ \log_2 \sqrt{3}, \ \log_2 9, \ \log_2 6$.

(b) $\log_4 3, \ \log_2 2\sqrt{3}, \ \log_2 \sqrt{3}, \ 2\log_9 3$.

6. Deduce the following identities from the other relations listed in Sect.2, for $a > 0, \ b > 1$:

(a) $a^0 = 1$.

(b) $\log_b 1 = 0$.

(c) $b^{\log_b y} = y$.

(d) $\log_b a = \dfrac{1}{\log_a b}$.

(e) $\sqrt{a} = a^{1/2}$.

7. Solve each of these equations for x exactly or correct to two decimal places:

(a) $2e^x + e^{x+3} - e^{2x} = 5$.

(b) $\dfrac{2^{x-1} - 3 \cdot 4^{x-2}}{4^x - 1} = 3$.

(c) $3(5^{1-x}) - 9\sqrt{5^{x-2}} = 2$.

(d) $(\lg x)^2 + \lg(\sqrt{x}) - 0.1 = 0$.

(e) $\log_2(2^{x-1} - 3 \cdot 4^x) = 2x$.

(f) $\ln \left| \dfrac{\sqrt{x} + \sqrt{3}}{\sqrt{x} - \sqrt{3}} \right| + \dfrac{1}{2} \ln \dfrac{1}{|x|} = 0$.

(g) $\lg \sqrt{x^2 - 3x + 2} = 1/\log_x 10$.

(h) $\log_x \log_x \log_x 2^{3x} = 0$.

8. The period, T, of a simple pendulum is the time for one complete cycle. It is given by the formula $T = 2\pi\sqrt{\frac{l}{g}}$ where l is the length of the pendulum and g the acceleration due to gravity. By taking logarithms of that equation, produce a linear equation in new variables from which you may determine g if you had experimental data on the variation of T with l.

9. Solve the pair of simultaneous equations
$27^x \; 3^{5y} = 9$ and $\log_2(1 - x) - 1 = \log_2 y$.

10. Without explicitly solving the equation $\sqrt{3 - x} + \sqrt{5x - 15} = 2$, explain why it cannot have a real solution.

11. Solve each of the following equations for x:

  (a) $\sqrt{4 - x} + \sqrt{4 + x} = \sqrt{16 - x^2}$.

  (b) $\sqrt{x - 1} + \sqrt{x - 4} = \sqrt{15 - 2x}$.

  (c) $\sqrt{2x^2 - x - 3} - \sqrt{x^2 + 2} = 0$.

  (d) $\sqrt{x^2 - 4x - 5} + \sqrt{x^2 - 2x - 3} = 2x + 2$.

12. Re-write each expression below as $a + b\sqrt{3}$ with a, b rational.

  (a) $\sqrt{125}\left(\dfrac{1}{\sqrt{60}} - \sqrt{\dfrac{96}{10}}\right)$.   (c) $\sqrt{4 - 2\sqrt{3}}$.

  (b) $\sqrt{\dfrac{1}{28}} \times \sqrt{\dfrac{105}{2}} \times \sqrt{\dfrac{5}{2}}$.

13. A circle has area $2\pi(2 + \sqrt{3})$. Without using a calculator, determine its radius in the form $a + b\sqrt{3}$ where a, b are rational numbers.

14. Reduce $\dfrac{3\sqrt{2} + 5}{(\sqrt{2} - 1)^3}$ exactly to its simplest form.

15. A bank pays 3% compound interest per month on its fixed deposits. If a lady deposits $\$\,10,000$, how much would she obtain when she makes a withdrawal three months later? (Note: Compounding at that rate means that at the end of each month the

total amount would be 1.03 times the amount at the start of that month). Calculate using two methods:
(a) The binomial theorem.
(b) Directly, using a calculator.

16. Find the coefficient of the x^4 term in the expansion of each of the following expressions:

(a) $\left(1 + \dfrac{x}{2}\right)^{10}$.

(c) $\left(x + \dfrac{1}{x}\right)^8$.

(b) $(1 + 2x^2)^{11}$.

(d) $(1 + 3x)^2 (2 + x)^6$.

17. Find the terms independent of x in the expansion of each expression in the previous question.

18. Using an appropriate sketch, show that the equation $e^x - 2\ln(x+1) = 1$ has exactly two solutions. Estimate the value of one of the solutions using an accurate plot.

19. Using an appropriate sketch, show that the equation $e^{-x} - \sin x = 0$ has exactly two solutions between $0 \leq x \leq \pi$ and no solutions between $\pi \leq x \leq 2\pi$. Estimate the value of the solution in the range $0 \leq x \leq \pi/2$.

20. Explain why the curve $y = \log_b x$ is a reflection of the curve $y = b^x$ about the line $y = x$.

1.6   Challenges

1. (a) Justify the identities involving exponents in Sect.(2) for integer exponents and then use those to motivate their extension to rational exponents.
(b) Derive the identities involving logarithms starting from those for exponents.
(c) Deduce $a^{-1} = \dfrac{1}{a}$ assuming the other identities for exponents.

2. Prove the following identities starting from the other relations involving logarithms that are listed in Sect.(2):

    (a) $P^{\log_b Q} = Q^{\log_b P}$.

    (b) $\log_b P = \dfrac{\log_a P}{\log_a b}$.

3. (a) Given the natural logarithm of the first ten integers, is it possible to evaluate, relatively quickly, the natural logarithm of any other positive integer, by hand, correct to one decimal place?

4. Prove that $\log_2 3$ cannot be a rational number; that is, it cannot be written as a/b for integers a, b.

5. (a) Rewrite $\dfrac{6^{2/3}}{3^{1/3} - 2^{1/3}}$ as $a + 12^{2/3} + 18^{2/3}$ with a an integer.

    (b) Reduce $\log_4 \left(\sqrt{3 + \sqrt{5}} - \sqrt{3 - \sqrt{5}} \right)$ to its simplest form.

1.7   Inquire and Investigate

1. Study the concept of "significant figures", as used in describing the precision of a quantity, and discuss how it differs from the use of "decimal places".

2. (a) Why are natural logarithms "natural"? Investigate their properties.
(b) Study the series representation of the exponential function e^x. (See C7 in the Supplementary chapter.)

3. The logarithmic spiral is the name of a specific curve. Look up its defining equation and investigate its unique mathematical properties. Are there naturally occurring phenomena described by such a spiral?

4. (a) Other than the Richter and decibel scales, what other scales in science are logarithmic?
(b) Read up on one these laws: Zipf, Pareto, Benford.

5. (a) Does the quantity 0^0 have any meaning? Explain.
(b) How many possible solutions are there to the equation $x^n = 1$ for n a positive integer? What are they? How many would be real?

(c) Is it possible to define the logarithm of a negative number? Would it be unique?

6. (a) What is the origin of the names "surd" and "logarithm"?
(b) How did the use of logarithms speed up multiplication before calculators were invented?
(c) Learn how slide-rules or "log tables", two essentially obsolete calculational aids, work.

Rest your eyes
Very long periods of close reading, especially under poor lighting
conditions, can strain the eyes and lead to several problems. Do take
regular breaks, look away from your book or computer screen, blink
your eyes and look out into the distance, preferably out of the window.

Chapter 2

Polynomials and Rational Functions

As the sun eclipses the stars by its brilliancy,
so the man of knowledge will eclipse the fame of others
in assemblies of the people if he proposes algebraic problems,
and still more if he solves them.
Brahmagupta

2.1   Introduction

Your friend Neo is standing on a platform which is at a height y_0 above ground: He throws a ball vertically upwards at time $t = 0$, with initial speed u. Ignoring air-resistance, what height y, above ground, would the ball reach in time t?

Well, in a physics class you would learn that $y(t) = y_0 + ut - \frac{gt^2}{2}$ where g is the acceleration due to gravity. The expression for $y(t)$ is an example of a quadratic polynomial. If Neo were interested to know when the ball would reach the ground, he would have to solve the equation $y_0 + ut - \frac{gt^2}{2} = 0$ which would have two solutions for t; however since t needs to be positive in this example, only one solution is selected, see exercise E(2).

The above example illustrates one occurrence of **quadratic equa-**

tions in science and the interest in their solution. The quadratic equation is a **polynomial equation** of **degree** 2.

In polynomial equations, $P(x) = 0$, the expression $P(x)$ is a sum of terms of the form $a_n x^n$, with n a non-negative integer and the a_n representing constants. The degree is the largest power of the variable appearing in the equation. For example, the linear equation $f(x) = 2x + 3$ is of degree one since the highest power of the variable x that appears is one.

In general, polynomial equations of degree larger than 2 are difficult to solve analytically, but sometimes, if one can guess one root, they can be reduced by division to lower degree polynomials.

If $P(x)$ is a polynomial with $P(\alpha) = 0$, then $x = \alpha$ is a **root** of that equation and one may **factorise** $P(x) \equiv (x - \alpha)Q(x)$ where $Q(x)$ is a polynomial of one degree lower; this is the **Factorisation Theorem**. The process of factorisation can continue if there are more roots.

On the other hand, if $x = \alpha$ is not a root of $P(x)$ then one would obtain a remainder when $P(x)$ is divided by $(x - \alpha)$ using the process of long division, $P(x) \equiv (x - \alpha)Q(x) + R$ with $P(\alpha) = R$ the remainder: This summarises the **Remainder Theorem**, which obviously generalises the Factorisation Theorem.

More generally, if the polynomial $P(x)$ is divided by another polynomial $D(x)$ of lower degree, then $P(x) \equiv D(x)Q(x) + R(x)$ where the degree of $R(x)$ is less than that of $P(x)$.

Brahmagupta

An Indian mathematician and astronomer who lived around 600 C.E., Brahmagupta is believed to be the first to define "zero" as a number and to state the arithmetical rules involving it with the positive and negative numbers. Among his many contributions, he described how to solve linear and quadratic equations, and is credited with a formula giving the area of a quadrilateral inscribed in a circle: If the sides are a, b, c, d, then
$A = \sqrt{(s - a)(s - b)(s - c)(s - d)}$ where s is half the perimeter. A limiting case of this formula produces Heron's formula for the area of a triangle, see practice question (C2) in the Trigonometry Chapter.

2.1.1 Partial Fractions

A rational function is a ratio of two polynomials. For example, the distance of the image formed by a thin lens is given by the formula $v = \dfrac{fu}{u - f}$ where u is the object distance from the lens and f the constant focal length of the lens.

In some calculations, for example integration of rational functions (see also problem (P19)), it is convenient to re-write the ratio as the sum of rational functions whose denominators are the factors of the original denominator. For example in

$$\frac{3x^2 + 2x + 1}{x^2 - 1} \equiv 3 - \frac{1}{(x + 1)} + \frac{3}{(x - 1)}, \tag{2.1}$$

the expression on the left has been re-written in terms of its partial fractions (see details in the worked example below).

More generally, for a rational function $P(x)/Q(x)$, the first step in the decomposition is to use long division to reduce the degree of the numerator to below that of the denominator. Next, each factor of $(x-a)$ in $Q(x)$ would require a partial fraction $A/(x-a)$. If the factor is repeated in Q, for example $(x - a)^2$, then one uses two partial fractions $A_1/(x - a)$ and $A_2/(x - a)^2$ for that factor. If Q contains a term that cannot be factorised (using real numbers), for example $x^2 + x + 1$, then the partial fraction for that term is of the form $(Ax + B)/(x^2 + x + 1)$, that is, the numerator is one degree lower than the denominator.

2.2 Relations and Properties

Given a quadratic equation $ax^2 + bx + c = 0$, its roots may be found by completing the square: Multiply the equation by $4a$, add b^2 to both sides and rearrange to get $4a^2x^2 + 4abx + b^2 = b^2 - 4ac$, or $(2ax + b)^2 = \Delta$ where

$$\Delta \equiv b^2 - 4ac \tag{2.2}$$

is called the **discriminant**. The solutions are therefore given by

$$x = \frac{-b \pm \sqrt{\Delta}}{2a}. \tag{2.3}$$

The two roots are real if and only if $\Delta \geq 0$; the case $\Delta = 0$ corresponds to a repeated root.

A different way of completing the square is often useful:
$$ax^2 + bx + c = a\left(x^2 + \frac{b}{a}x + \frac{c}{a}\right) = a\left(x + \frac{b}{2a}\right)^2 + \left(c - \frac{b^2}{4a}\right).$$
This form shows that the quadratic equation has its extremal value at $x = -b/(2a)$, the point being a maximum if $a < 0$ and a minimum if $a > 0$ (this is discussed more below).

One may obtain various relations between the roots of a quadratic equation without explicitly solving for its roots. Denoting the two roots of a quadratic equation by α and β, then by the Factorisation Theorem $ax^2 + bx + c \equiv a(x - \alpha)(x - \beta)$, which on expanding the right side and comparing coefficients leads to the two basic relations

$$
\begin{aligned}
\alpha + \beta &= \frac{-b}{a}, \\
\alpha\beta &= \frac{c}{a}.
\end{aligned}
\tag{2.4}
$$

Other relations may be derived from these, for example

$$
\begin{aligned}
\alpha - \beta &= \pm\sqrt{(\alpha - \beta)^2} \\
&= \pm\sqrt{(\alpha + \beta)^2 - 4\alpha\beta},
\end{aligned}
\tag{2.5}
$$

and

$$\alpha^2 + \beta^2 = (\alpha + \beta)^2 - 2\alpha\beta. \tag{2.6}$$

What does the curve $y = ax^2 + bx + c$ look like? It is determined by the sign of a: When $a > 0$, we see that y is positive and increasing for very large values of $|x|$, so the curve must have a minimum point. Next, by completing the square, or equivalently by looking at the discriminant, we can deduce if the curve has any intersection with the $y = 0$ axis (that is, any real roots). Similarly, the case $a < 0$ leads to a "hump" shaped curve which reaches a maximum value.

For a cubic curve $y = ax^3 + bx^2 + cx + d$, the sign of the leading coefficient again determines its the main shape. If $a > 0$, the curve rises upwards for large positive x and decreases for large negative x.

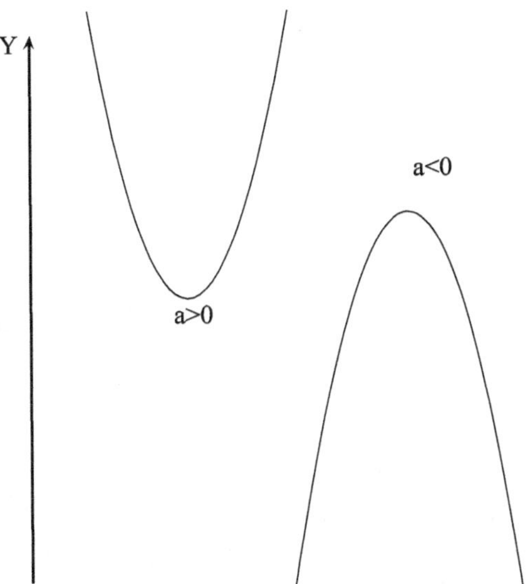

Figure 2.1: The shape of the parabola $y = ax^2 + bx + c$.

In between, it will either have a local minimum and a local maximum or gently undulate. We leave it to you to figure out the situation for $a < 0$.

In the differential calculus chapter you will learn how to reach the conclusions of the last two paragraphs using other techniques.

2.3 Worked Examples

1. *If the function $g(x) = ax^2 + bx + c$ is always positive for any real x, what can you say about the following:*
   (a) The sign of a; (b) The value of the discriminant?

   Solution:
   (a) Since the leading term of a polynomial dominates its value for large values of $|x|$, a must be positive if $g(x)$ is always positive.

   (b) However while $a > 0$ is necessary, it is not sufficient to guarantee $g(x) > 0$ for all x; we also need to make sure $g(x)$ stays above the x-axis for lower values of x, in other words it has no real roots. So we need $\Delta \equiv b^2 - 4ac < 0$.

Answer: (a) $a > 0$, (b) $b^2 - 4ac < 0$.

---

2. *For which values of x is $13 - 4x > (x - 4)^2$?*

Solution:
Bringing all the terms to one side gives the simpler condition $x^2 - 4x + 3 < 0$. So we need to find the region of the curve $f(x) = x^2 - 4x + 3$ below the x-axis. The roots of $f(x) = 0$ are $x = 1$ and $x = 3$ and from a sketch of the curve we deduce that the required region is $1 < x < 3$.

Answer: $1 < x < 3$.

---

3. *If the two expressions for a polynomial, $x^2 - 5x + 7$ and $A(x - 1)(x - 2) + B(x - 2)^2 + C$, are identical, determine the constants A, B, C.*

Solution:
If two expressions for a polynomial are identical, it means that that they both give the same answer for each value of the variable x. There are two ways to approach this question. One method is to expand the second expression and write it in descending powers of x; then by comparing coefficients of equal powers of x in the two expressions we can deduce the values of the constants.

A more efficient method is to recognise that the second expression simplifies for several values of x and to compare the two expressions at those values. Indeed, choosing $x = 2$ gives us $2^2 - 5(2) + 7 = 0 + 0 + C$, so $C = 1$. Then choosing $x = 1$ gives $1 - 5 + 7 = B + C$, hence $B = 2$. Finally to determine A you can choose any value of x not already used; choosing $x = 0$ gives $7 = 2A + 4B + C$, so $A = -1$.

We encourage you to re-do this exercise using the first method mentioned and compare the answers.

Answer: $A = -1$, $B = 2$, $C = 1$.

---

4. *Solve for all the roots of the equation $f(x) = x^3 + 2x^2 - 1 = 0$.*

Solution:
We first attempt to find one root by trial (from question (C2) we are led to try $x = \pm 1$). We find $f(-1) = 0$ and so $(x + 1)$ is one factor:
$f(x) = (x + 1)Q(x)$ where $Q(x)$ may be found by division,

$$Q(x) \;=\; \frac{x^3 + 2x^2 + 0x - 1}{x + 1} \;.$$

Notice that we have inserted $0x$ as a place holder for the absent x coefficient; you may find it useful to do likewise to avoid confusion or mistakes when doing the long division below. (The 0 value placed here plays a role similar to the one where we use it to distinguish the number 1201 from 121 in our usual decimal notation; recall that 1201 is $1 \cdot 10^3 + 2 \cdot 10^2 + 0 \cdot 10^1 + 1 \cdot 10^0$).

Here is the explicit long-division[1]:

$$x + 1 \overline{) \; x^3 + 2x^2 \qquad - 1} \quad \Big| \; x^2 + x - 1 \tag{2.7}$$

$$
\begin{array}{r}
x^2 + x - 1 \\
\hline
x + 1 \,) \quad x^3 + 2x^2 \qquad\;\; - 1 \\
-\,x^3 \;\; -\, x^2 \\
\hline
x^2 \\
-\,x^2 - x \\
\hline
-\,x - 1 \\
x + 1 \\
\hline
0
\end{array}
$$

So $f(x) = (x+1)(x^2 + x - 1) = 0$ and the other roots of $f(x)$ are the roots of the quadratic which you should find and compare with our answer.

Answer: $x = -1$, $(-1 \pm \sqrt{5})/2$.

---

5. Obtain the right-hand-side of eq.(2.1) from the left.

Solution:
The first step is to use long division to reduce the degree of the numerator below that of the denominator,

$$
\begin{array}{r}
3 \\
\hline
x^2 - 1 \,) \quad 3x^2 + 2x + 1 \\
-\,3x^2 \qquad\;\; + 3 \\
\hline
2x + 4
\end{array}
\tag{2.8}
$$

So

$$\frac{3x^2 + 2x + 1}{x^2 - 1} \;\equiv\; 3 + \frac{2x + 4}{(x+1)(x-1)}.$$

---

[1] Here the minus signs have been inserted at the relevant points so one is adding sums; alternatively you can exclude those minus signs and remember to subtract quantities.

The fractional part may now be split into partial fractions. We expect

$$\frac{2x+4}{(x+1)(x-1)} \equiv \frac{A}{(x-1)} + \frac{B}{(x+1)}$$

which on cross-multiplication gives

$$2x+4 \equiv A(x+1) + B(x-1).$$

You may now deduce, as in worked example 2, that $A=3$, $B=-1$.

Answer: Displayed in (2.1).

---

6. *The polynomial $g(x) = x^4 - px^3 - qx^2 + 50x + 75$ is divisible by $x^2 - 25$.*

   (a) Determine the constants p, q and factorise the polynomial completely.

   (b) Find the remainder when the polynomial is divided by $(x-2)$.

   (c) Find the remainder when the polynomial is divided by $x^2 - 3x + 2$.

   (d) Using the results of part (a), solve the equation
   $(2x-1)^4 - 2(2x-1)^3 - 7(4x-2)^2 + 100x + 25 = 0.$

Solution:

(a) Since $x^2 - 25 = (x-5)(x+5)$, we get two conditions from the Factorisation Theorem, $g(5) = 0$ and $g(-5) = 0$, which you can use to solve for $p = 2$ and $q = 28$. Then by long division,

$$\frac{x^4 - 2x^3 - 28x^2 + 50x + 75}{x^2 + 0x - 25} = x^2 - 2x - 3 = (x-3)(x+1),$$

so $g(x) = x^4 - 2x^3 - 28x^2 + 50x + 75 = (x+1)(x-3)(x-5)(x+5)$.

(b) Since $(x-2)$ is not a factor of $g(x)$, dividing the latter by the former would give a remainder which can easily be computed by the Remainder Theorem to be $g(2) = 63$.

(c) Use long division to verify that

$$x^4 - 2x^3 - 28x^2 + 50x + 75$$
$$= (x^2 + x - 27)(x^2 - 3x + 2) + (129 - 33x).$$

Hence the remainder is $(129 - 33x)$.

(d) We need to solve
$(2x - 1)^4 - 2(2x - 1)^3 - 7(4x - 2)^2 + 100x + 25 = 0$, using the hint that it is related to the original equation $g(x) = 0$. You might notice that the first two terms of the two equations have a similar structure (with x replaced by $2x - 1$ in the second), so let us try to re-write the second equation so that all its terms have a structure similar to that of $g(x)$:

$$(2x - 1)^4 - 2(2x - 1)^3 - 7(4x - 2)^2 + 100x + 25$$
$$\equiv (2x - 1)^4 - 2(2x - 1)^3 - 28(2x - 1)^2 + 50(2x - 1) + 75$$
$$\equiv g(y),$$

where we have set $2x - 1 = y$. Thus we need to solve $g(y) = 0$, a quartic polynomial with at most four real roots. Since we know from part (a) that $g(5) = g(-5) = g(-1) = g(3) = 0$, we can set those roots equal to $y = 2x - 1$ to get the values of x which are roots of the new polynomial. Hence $x = 3, -2, 0, 2$.

Answers: (a) $g(x) = (x + 1)(x - 3)(x - 5)(x + 5)$, (b) 63,
(c) $(129 - 33x)$, (d) $x = 3, -2, 0, 2$.

---

7. *(a) If the roots of the quadratic equation $2x^2 - 7x + p = 0$ are α and β,*
(a) Construct a second quadratic equation with roots $\alpha^2 - 1/\beta^2$, and $\beta^2 - 1/\alpha^2$.
(b) For the original quadratic equation, determine the value of p for which the equation has no real roots.
(c) For the second quadratic equation, determine the value of p

for which its roots are equal.

Solution:

(a) Using eq.(2.4) we get $\alpha\beta = p/2$ and $\alpha + \beta = 7/2$. Denote the roots of the new equation by $\alpha' = \alpha^2 - 1/\beta^2$ and $\beta' = \beta^2 - 1/\alpha^2$. Then the new equation is $a(x - \alpha')(x - \beta') = 0$. Since the constant a does not affect the subsequent analysis, we set $a = 1$ in what follows. Then expanding, the new equation is $x^2 - (\alpha' + \beta')x + \alpha'\beta' = 0$. Thus we need explicit expressions for $(\alpha' + \beta')$ and $\alpha'\beta'$. We have

$$
\begin{aligned}
\alpha' + \beta' &= \alpha^2 - \frac{1}{\beta^2} + \beta^2 - \frac{1}{\alpha^2} \\
&= \alpha^2 + \beta^2 - \left(\frac{1}{\alpha^2} + \frac{1}{\beta^2}\right) \\
&= (\alpha^2 + \beta^2)\left(1 - \frac{1}{\alpha^2\beta^2}\right) \\
&= ((\alpha + \beta)^2 - 2\alpha\beta)\left(1 - \frac{1}{\alpha^2\beta^2}\right) \\
&= \left((\tfrac{7}{2})^2 - p\right)\left(1 - (\tfrac{2}{p})^2\right)
\end{aligned}
\tag{2.9}
$$

and

$$
\begin{aligned}
\alpha'\beta' &= \left(\alpha^2 - \frac{1}{\beta^2}\right)\left(\beta^2 - \frac{1}{\alpha^2}\right) \\
&= \alpha^2\beta^2 - 2 + \left(\frac{1}{\alpha^2\beta^2}\right) \\
&= \left((\tfrac{p}{2})^2 + (\tfrac{2}{p})^2\right) - 2
\end{aligned}
\tag{2.10}
$$

which fix the equation $x^2 - (\alpha' + \beta')x + \alpha'\beta' = 0$.

(b) For there to be no real roots, we need the discriminant to be negative: $7^2 - 4(2)p < 0 \Rightarrow p > 49/8$.

(c) For the new equation to have equal roots we need $\alpha' = \beta'$ which implies $\alpha^2 - 1/\beta^2 = \beta^2 - 1/\alpha^2 \Rightarrow$

$$
0 = \alpha^2 - \beta^2 + \left(\frac{1}{\alpha^2} - \frac{1}{\beta^2}\right)
$$

$$
\begin{aligned}
0 &= (\alpha^2 - \beta^2)\left(1 - \frac{1}{\alpha^2\beta^2}\right) \\
&= (\alpha - \beta)(\alpha + \beta)\left(1 - \frac{1}{\alpha^2\beta^2}\right) \\
&= \pm\sqrt{(\alpha + \beta)^2 - 4\alpha\beta}\;(\alpha + \beta)\left(1 - \frac{1}{\alpha^2\beta^2}\right) \\
&= \pm\sqrt{(\tfrac{7}{2})^2 - 2p}\;\left(\tfrac{7}{2}\right)\left(1 - (\tfrac{2}{p})^2\right),
\end{aligned}
\tag{2.11}
$$

which has the solutions $p = 49/8$ and $p = \pm2$.

Answers: (a) As described above, (b) $p > 49/8$,
(c) $p = 49/8,\ \pm2$.

2.4   Exercises

1. A line segment is divided into two parts of length a and b such that $\dfrac{a}{b} = \dfrac{a+b}{a} \equiv \phi$.

   (a) Determine the numerical value of ϕ, known as the "Golden Ratio".

   (b) Optional: Study some properties and applications of the Golden Ratio.

2. (a) In Sect.(1), Neo was interested in the time it would take the ball to reach the ground. Determine that time if $y_0 = 10$ m, $u = 5$ m/s and $g = 10$ m/s^2.

   (b) By writing y in the form $y = a(t + b)^2 + c$, determine the maximum height reached by the thrown ball and the time when it reaches that maximum height.

   (c) For how long did the ball stay 5 m or more above ground?

   (d) If Neo were to throw the ball the same way on the Moon, where the g value is 1/6-th of that on Earth, what would the answers to parts (a, b) and (c) be?

3. Solve the following equations for x:

(a) $2x + 2\sqrt{x} - 1 = 0$.

(b) $1 + \dfrac{3}{\sqrt{x}} - \dfrac{1}{x} = 0$.

(c) $\dfrac{1}{1-x} + \dfrac{1}{x+2} = 3$.

4. Find the range of values of x for which:

(a) $2 \geq 5x + x^2$.

(b) $4x \geq (x-1)(x-3)$.

(c) $2x + 5 \leq (3-x)^2$.

5. The roots of the quadratic equation $2x^2 - 7x + 1 = 0$ are α and β.

(a) Determine the values of $\alpha + \beta$, $\alpha\beta$ and $1/\alpha + 1/\beta$ without solving the quadratic equation.

(b) Solve the quadratic equation to determine α, β explicitly and verify your answers to part (a).

6. If $f(x) = x^2 + bx + c$, determine the constants b, c for the separate exercises below and sketch the curve $y = f(x)$:

(a) $f(x)$ has a double root at $x = 5$.

(b) $f(x)$ has roots at $x = -1$ and $x = 2$.

7. For each $f(x)$ determined in the previous question, sketch $y = |f(x)|$ and determine the range that y takes for x between 0 and the positive root.

8. If $f(x) = x^2 + bx + 2b$, determine constraints on the constant b for the separate exercises below:

(a) $f(x)$ has no real roots.

(b) $f(x)$ has a root at $x = -1$.

(c) $f(x)$ has one positive and one negative real root.

(d) $f(x)$ has two negative real roots.

9. A student is told that the roots of a quadratic equation $f(x) = 0$ satisfy $\alpha + \beta = 2$ and $\alpha\beta = -2$.

(a) Determine the roots.

(b) Determine the expression for $f(x)$ if the coefficient of the x^2 term is 3.

(c) If instead of the information in part (b), you were told that the coefficient of the x^2 term is a and the quadratic equation $y = f(x)$ meets the line $y = x - 3a$ at only one point, can you determine $f(x)$ uniquely?

10. The cubic equation $f(x) = x^3 - ax^2 - x + 3$ has a root at $x = -1$.
(a) Factorise $f(x)$.
(b) Sketch the equation $y = f(x)$.
(c) For what values of c will the curve in part (b) intersect the curve $x^3 - y - c = 0$?

11. Re-write each of the following in terms of partial fractions.

(a) $\dfrac{2x + 1}{x^2 + 3x + 2}$.

(b) $\dfrac{5x^2 + 2x + 1}{x^2 + 3x + 2}$.

(c) $\dfrac{5x^2 + 2x + 1}{x + 1}$.

2.5   Problems

1. The A series of paper sizes start with the largest, $A0$, having an area $1m^2$. If the sides of the (rectangular) $A0$ sheet are denoted by x and y, with x the longer side, then the $A1$ sheet is obtained by dividing the $A0$ sheet in half along its longer side, obtaining dimensions $x/2$ by y, with y now the longest side. If the aspect ratios of the two sizes are equal, that is, $\dfrac{x}{y} = \dfrac{y}{(x/2)}$,
(a) Determine x and y.
(b) If succeeding paper sizes ($A2, A3, A4...$) are defined by dividing each previous size by two along the longer side, determine the dimensions of the $A4$ sheet.
(c) Compare your answer in part (b) with a direct measurement of an $A4$ sheet.

2. If the function $g(x) = ax^2 + b^2x + c$, defined over the whole real line, is always less than 3 for any real x, and has a root at $x = 0$, what can you say about the following:
(a) The sign of a.

(b) The value of c.
(c) The sign of the other root.

3. Bjorg, a sports scientist, uses a high-speed camera to study the trajectories of tennis balls shot from a tennis ball cannon. He finds that the trajectory is well modelled by $y = ax - bx^2$ with the origin, $(x = 0, y = 0)$, at the mouth of the cannon and the ground at $y = -12.5$. He also finds that a shot ball reaches a maximum height 112.5 above ground and hits the ground at $x = 25$.
(a) Determine the values of a and b (both positive) and sketch the trajectory.
(b) What are the coordinates of the maximum point in the trajectory?
(c) At what value of x does the shot ball reach $y = 0$ again?
(d) For which range of x values does the ball satisfy $y \geq 50$?

4. Solve the following equations for x:

(a) $\dfrac{1}{5 - x^3} + \dfrac{1}{5 + x^3} \leq 5.$
(b) $\dfrac{1}{x^2 + 3} + \dfrac{1}{x^2 - 3} \geq \dfrac{-2x^2}{11}.$

5. Solve each of the following pair of simultaneous equations:
(a) $y = 2x^2 - 3x + 3$ and $y = x + 2$.
(b) $y = x^2 - x - 1$ and $y = 3x^2 - 4$.
(c) $y^2 = 2x^2 - 3x + 3$ and $y = -x + 2$.

6. The quadratic equation $ax^2 + bx + c = 0$ has roots α and β, with $\gamma \equiv \alpha/\beta > 1$. Find the value of $\dfrac{\gamma - 1}{\gamma + 1}$ in terms of a, b, c.

7. If the roots of the quadratic equation $2x^2 - 7x + p = 0$ are α and β, with $\alpha > \beta$, determine the following in terms of p:
(a) $\alpha - \beta$.
(b) $(\beta/\alpha + \alpha/\beta)$.
(c) $\alpha^4 + \beta^4$.

8. A student mistakenly studies the equation $g(x) = cx^2 + bx + a = 0$ instead of the equation $f(x) = ax^2 + bx + c = 0$ his teacher intended.

(a) If the roots of $g(x)$ are α' and β', while those of $f(x)$ are α and β, determine $\alpha' + \beta'$ and $\alpha'\beta'$ in terms of α and β.
(b) Do the curves $y = g(x)$ and $y = f(x)$ ever intersect? Determine when.

9. A rectangle has sides of length a and $1/a$. If the length of the diagonal is $3/8$ times the length of the perimeter, determine a either approximately, correct to two decimal places, or exactly.

10. Assuming that the infinite sum represented by
$$x = \sqrt{2 + \sqrt{2 + \sqrt{2 + \ldots}}} \quad \text{exists, determine the value of } x.$$
(Hint: Square both sides).

11. If a spherical tank of radius R contains a volume V of fuel when it is filled to a height h $(\leq 2R)$, it can be shown that
$$V = \frac{\pi h^2}{3}(3R - h).$$
(a) Verify that the formula is correct when $h = 2R$.
(b) If $R = 10$ and $V = 29\pi/3$, find the cubic equation that determines h and find one of its roots by trial.
(c) Find all roots of the equation in part(b) and discuss which are the possible physical solutions.
(d) Optional: Use calculus to prove the formula relating V to h.

12. You are given a square piece of cardboard of side a to construct an open top box as follows: First cut out squares of dimensions x by x at each corner of the cardboard, then fold the cardboard to form a box of dimensions x by $(a - 2x)$ by $(a - 2x)$.
(a) Find the values of x which make the volume of the box a minimum and explain the physical situation those values correspond to.
(b) Hence, or otherwise, explain why for some value of x the volume of the box would be a maximum.
(c) If $a = 1$, estimate the maximum volume of the box using an appropriate plot.
(d) Optional: Solve for part(c) using calculus.

13. A teacher tells her class that the quartic polynomial
$P(x) = x^4 + ax^3 + bx^2 + 3$ is identical to the expression

$S(x) = x(x+1)(x-2)Q(x) + c + 1$ written by a student.
(a) Determine the constants a, b and c.
(b) Hence find an expression for $Q(x)$.
(c) What is the remainder when the polynomial $P(x)$ is divided by $x - 1$?

14. Determine the real roots of the equation $x^3 + 1 = 0$. Hence factorise the expression $a^3 + b^3$.

15. The cubic polynomial equation $g(x) = 0$ has roots 1, p and $2p$ and the coefficient of x^0 in $g(x)$ is 5. If $g(x)$ leaves a remainder of -2 when divided by $(x - 2)$, determine the possible values of p.

16. The Van der Waals equation for a gas may be written in the cubic form

$$v^3 - \frac{1}{3}\left(1 + \frac{8T}{p}\right)v^2 + \frac{3}{p}v - \frac{1}{p} = 0 \qquad (2.12)$$

where p, v, T are known as the reduced pressure, volume and temperature.
(a) Solve the equation for v when $p = T = 1$ (these are called the "critical values").
(b) Show that equation (2.12) may also be written in the usual form $\left(p + \frac{3}{v^2}\right)\left(v - \frac{1}{3}\right) = \frac{8T}{3}$.
(c) Assuming T is constant, re-write equation (2.12), expressing p in terms of v, in partial fraction form.
(d) Optional: What is the application of the Van der Waals equation?

17. Re-write each of the following in terms of partial fractions.

(a) $\dfrac{x^4 + 1}{x^3 + x}$.

(b) $\dfrac{5x^2 + 2x + 1}{x^2 + 2x + 1}$.

(c) $\dfrac{5x^2 + 2x + 1}{(x^2 + 2x + 1)(x + 2)}$.

(d) $\dfrac{5x^2 + 2x + 1}{(x^2 + 1)(x + 2)}$.

18. Since for $c \neq 0$, $x = 0$ is not a root of $ax^2 + bx + c = 0$, the quadratic equation may be written as $ax + b = \dfrac{-c}{x}$; the latter form representing the intersection of a line with a hyperbola. Taking $a > 0$ for simplicity, use the latter form to deduce the condition when the quadratic equation has no real roots. Compare with the usual condition.

19. Evaluate $\displaystyle\sum_{k=2}^{\infty} \frac{1}{k^2 - 1}$ using partial fractions.

2.6   Challenges

1. Show that the sum $\displaystyle\sum_{k=2}^{\infty} \frac{1}{k+1}$ diverges. How then can you justify the manipulations leading to the result in question (P19)?

2. Consider the cubic equation $x^3 + bx^2 + cx + d = 0$, with integral coefficients b, c and d.
   (a) Prove that its real solutions are either irrational or integers.
   (b) Show that if integral solutions exist, they must be factors of d.
   (c) Hence find all solutions of $x^3 + 2x^2 - 5x - 10 = 0$.
   (d) Generalise the proof in part (a) to cubic equations with leading term ax^3, with a integral.

3. Sketch the general form of the cubic equation $x^3 + bx^2 + cx + d = 0$. By determining its turning points, and the values at those points, deduce conditions for the cubic equation to have three real roots.

4. For $d \neq 0$, $x = 0$ is not a root of $x^3 + bx^2 + cx + d = 0$ and so the cubic equation may be re-written as $x^2 + bx + c = \frac{-d}{x}$. The latter form represents the intersection of a parabola with a hyperbola. Use the latter form to deduce the condition when the cubic equation has only one real root. Compare with the results of the previous question.

5. Solve for x in $\dfrac{1 + x + 2x^2}{3x} - \dfrac{x}{2x + 4x^2 + 2} = 7.$

2.7   Inquire and Investigate

1. Quadratic equations occur in chemistry when one wishes to solve for the equilibrium concentration of chemicals in a reaction. Study one example in detail.

2. (a) Is it possible for a quadratic equation with real coefficients to have only one real root? Explain.
   (b) Explain why the sum and products of roots of a quadratic equation (with real coefficients) are always real even if the equation has no real roots, for example $x^2 + x + 1 = 0$.

3. There are other ways to "complete the square" when solving the quadratic equation, and also different formulae for its roots. Investigate these different methods and discuss some applications of the alternatives.

4. (a) Study how to obtain the roots of a general cubic polynomial analytically.
   (b) Is it possible to also find explicit expressions for the roots of a quartic, quintic or higher degree polynomial?

5. Study an algorithm that can be used to find the roots of polynomial equations numerically. (See question (12) of Sect.(9.1) or question (11) of Sect.(9.2)).

6. What is "synthetic division"? How and why does it work?

Stretch
Relieve muscular tension by taking breaks: Stand up, slowly loosen your joints, and gently stretch. Be especially aware of your neck, shoulders and lower back.

Chapter 3

Simultaneous Equations and Matrices

To divide a cube into two other cubes, a fourth power or in general any power whatever into two powers of the same denomination above the second is impossible, and I have assuredly found an admirable proof of this, but the margin is too narrow to contain it.

Pierre Fermat

3.1 Introduction

Neo's sister Eona is standing at the foot of a flat platform which is inclined at an angle α to the ground. She throws a ball up the platform with initial speed u and making an angle β with the platform. Ignoring air-resistance, where would the ball hit the platform?

The trajectory of the ball is given by a parabola, while the platform may be modelled by a straight line. So the answer to the above question requires us to find the intersection point of the parabola and straight line, an example of solving **simultaneous equations**.

For simple cases, two simultaneous equations may be reduced to one equation by eliminating one common variable between them. However even when this is possible, the resulting equation might be too difficult to solve analytically, as in the example of the two equations $y = 2e^{-x}$ and $y - x - 1 = 0$. For such cases one may find an approximate solution using **graphical methods**: Plot the two curves $y = 2e^{-x}$ and $y = x+1$ on the same graph and see where they intersect; this is illustrated in one of the worked examples.

If the simultaneous equations are all **linear**, then a systematic procedure to solve them exactly uses **matrices**, as discussed in the next section.

What about inequalities? If there is a combination of equations and inequalities, then typically the boundaries of the solution set are first determined by treating all of them as equalities. For example $y = x^2$ and $y < x - 1$ is equivalent to $x - 1 > x^2$ or $x^2 - x + 1 < 0$. Solving the equation $x^2 - x + 1 = 0$ and making a sketch will help determine the solution set for the original problem involving the inequality.

Fermat

Pierre de Fermat was a lawyer and amateur mathematician who lived in 17th century France. He made numerous contributions to mathematics, such as developing probability theory together with Pascal, and developing techniques that were similar to those of differential calculus (which were later developed fully by Newton and Leibniz). Fermat had a habit of communicating his results with little or no proof, leaving it to later mathematicians to fill in the gaps. His most famous statement came to be known as "Fermat's Last Theorem", which stated that there were no integral solutions (x, y, z) to the equation $x^n + y^n = z^n$ for integral $n > 2$. Fermat stated he had a proof which was too long to fit into the margin of a book by Diophantus that he was reading (apparently he had the habit of scribbling his proofs, if any, in the margins of books). Despite the intense efforts of dozens of mathematicians over a period of over 200 years, the Last Theorem defied proof until 1994 when Andrew Wiles, using sophisticated techniques, finally confirmed Fermat's claim.

3.1.1   Matrices

A matrix is an array of numbers which may be added and multiplied to other matrices according to certain rules. The manipulation of such matrices is useful for solving simultaneous linear equations and related problems.

A $m \times n$ matrix A has m rows and n columns and its elements may be denoted by a_{ij} where the first index $i = 1, 2, ..., m$ labels the rows and the second index $j = 1, 2, ..., n$ labels the columns. For example, here is a 2×3 matrix

$$A = \begin{pmatrix} 1 & 0 & -2 \\ 0.5 & 3 & 1.2 \end{pmatrix} \tag{3.1}$$

some of whose elements are $a_{11} = 1$, $a_{12} = 0$, $a_{21} = 0.5$, etc.

Two matrices, A and B, may be added only if both have the same number of rows, and both have the same number of columns; their sum, $A + B = C$ is simply a new matrix whose elements are the sum of the corresponding terms of the two matrices: $c_{ij} = a_{ij} + b_{ij}$.

Two matrices, A and B, may be multiplied together only if the number of columns of the first matrix equals the number of rows of the second matrix. For example, if A is a 2×3 matrix and B a 3×4 matrix then you may form the product AB which yields a 2×4 matrix according to the rules described below. In this example, it is not possible to form the product BA. More generally, a $m \times n$ matrix multiplying from the left a $n \times p$ matrix yields a $m \times p$ matrix[1].

How are the elements of $C = AB$ calculated for the matrices of the previous paragraph? We illustrate the rules through the c_{23} element of C; it is obtained by multiplying the second row (index 2) of A to the third column (index 3) of B term by term and adding the pieces. For example, if A is given by eq.(3.1) and B is the matrix

$$\begin{pmatrix} 2 & 1 & 3 & 1 \\ 5 & 2 & 8 & -1 \\ 1 & -2 & 1 & 0 \end{pmatrix} \tag{3.2}$$

then the product $C = AB$ is given by

$$\begin{pmatrix} 0 & 5 & 1 & 1 \\ 17.2 & 4.1 & 26.7 & -2.5 \end{pmatrix}. \tag{3.3}$$

---

[1]Unlike multiplication of ordinary numbers, matrix multiplication is non-commutative in general. That is, even if the matrix products AB and BA were possible, the result will in general not be the same matrix.

Check: The c_{23} value is $(0.5 \times 3) + (3 \times 8) + (1.2 \times 1) = 26.7$. More generally the c_{ij} element is obtained by multiplying the i-th row of A to the j-th column of B term by term and adding the pieces.

You might well ask why matrix multiplication is defined in this way. It turns out that this definition is natural and useful in many contexts as we shall immediately see.

Consider the pair of simultaneous equations in the variables (x, y):

$$ax + by \;=\; e \tag{3.4}$$
$$cx + dy \;=\; f\,, \tag{3.5}$$

with a, b, c, d, e, f constants. Graphically, the problem corresponds to the intersection of two lines.

In matrix notation these equations are written as

$$MX = R \tag{3.6}$$

where

$$M = \begin{pmatrix} a & b \\ c & d \end{pmatrix} \tag{3.7}$$

is the 2×2 matrix,

$$X = \begin{pmatrix} x \\ y \end{pmatrix} \tag{3.8}$$

is the column vector (a 2×1 matrix) representing the variables, and

$$R = \begin{pmatrix} e \\ f \end{pmatrix} \tag{3.9}$$

is another column vector containing some constants.

The **determinant** of the 2×2 matrix M is defined by

$$\det(M) = ad - bc\,. \tag{3.10}$$

If $\det(\mathrm{M}) \neq 0$ then one may define an **inverse matrix**

$$M^{-1} = \frac{1}{\det(\mathrm{M})} \begin{pmatrix} d & -b \\ -c & a \end{pmatrix} \tag{3.11}$$

and the solution to the problem (3.6) is

$$X = M^{-1}R = \frac{1}{\det(\mathrm{M})} \begin{pmatrix} de - bf \\ -ce + af \end{pmatrix},$$ (3.12)

where the right-hand-side has been obtained using the rules of matrix multiplication as defined earlier.

If $\det(\mathrm{M})= 0$ the matrix is termed **singular** and there is no inverse matrix; geometrically, the two lines do not intersect.

You can check that

$$MM^{-1} = M^{-1}M = \begin{pmatrix} 1 & 0 \\ 0 & 1 \end{pmatrix},$$ (3.13)

the last matrix being the **identity matrix**, usually denoted by the letter I.

We encourage you to re-obtain the solution (3.12) from (3.5) by using the usual process of eliminating one variable, solving for the other, and then finally deducing the first variable.

It is possible to use matrices to solve simultaneous equations for more than two variables but we will not do so in this text.

In addition to the practice questions below, there are questions involving simultaneous equations and inequalities in the other chapters which involve, for example, logarithms and trigonometric functions.

3.2   Worked Examples

1. *The* 200 *students of a school are teamed up into groups of either* 3 *or* 5. *If the total number of teams formed is* 44, *determine the number of teams with* 3 *members.*

   Solution:
   Let x and y denote respectively the number of teams with 3 or 5 members. Then we have

   $$\begin{aligned} 3x + 5y &= 200 \\ x + y &= 44. \end{aligned}$$

You can solve these equations by using the second equation to eliminate y from the first equation, getting the solution for x, and then y. Alternatively you can use matrix methods. The problem can be written as

$$\begin{pmatrix} 3 & 5 \\ 1 & 1 \end{pmatrix} \begin{pmatrix} x \\ y \end{pmatrix} = \begin{pmatrix} 200 \\ 44 \end{pmatrix}. \tag{3.14}$$

The determinant of the 2×2 matrix is $3 \times 1 - 5 \times 1 = -2$. The inverse matrix is therefore

$$\frac{-1}{2}\begin{pmatrix} 1 & -5 \\ -1 & 3 \end{pmatrix} \tag{3.15}$$

and the solution to the problem is given by

$$\begin{pmatrix} x \\ y \end{pmatrix} = \frac{-1}{2}\begin{pmatrix} 1 & -5 \\ -1 & 3 \end{pmatrix}\begin{pmatrix} 200 \\ 44 \end{pmatrix}$$

$$= \frac{-1}{2}\begin{pmatrix} 1 \times 200 - 5 \times 44 \\ -1 \times 200 + 3 \times 44 \end{pmatrix}$$

$$= \begin{pmatrix} 10 \\ 34 \end{pmatrix}.$$

Answer: 10.

---

2. *Find the range of values of the parameter k so that the two curves $2y - kx = 1$ and $y = 3x^2 + 2$:*
   (a) Do not intersect.
   (b) Intersect at only one point.
   (c) Find the solution set for part (b).
   (d) Explain the result in part (c) graphically.

Solution:
Since the first equation is linear and the second a quadratic, substituting y from the first into the second gives a quadratic

equation which can be analysed analytically:
$(1+kx)/2 = 3x^2+2 \Rightarrow 6x^2-kx+3 = 0 \Rightarrow x = (k\pm\sqrt{k^2-72})/12.$
So there are real solutions if and only if $k^2 \geq 72$.

(a) If the curves do not intersect, there will be no solutions to the last quadratic (and vice versa). This happens when $-\sqrt{72} < k < \sqrt{72}$.

(b) For a single solution, the roots of the quadratic must be identical. This happens when $k = \pm\sqrt{72}$.

(c) When $k^2 = 72$, $k = \pm\sqrt{72} = \pm6\sqrt{2}$ and $x = \pm1/\sqrt{2}$. That is, there is one x value for each possible sign of k and the corresponding y value is $y = (1 + kx)/2 = 7/2$.

(d) It is instructive to solve the problem graphically from the start (you should do it). It represents the intersection of a straight line with a parabola. The solution set in part (b) is then seen to consist of two tangent lines, and from the symmetry of the figure you can see why the y value is the same for the two x values.

Answers: (a) $-\sqrt{72} < k < \sqrt{72}$, (b) $k = \pm\sqrt{72}$, (c) $(1/\sqrt{2}, 7/2)$ and $(-1/\sqrt{2}, 7/2)$,
(d) The two solutions correspond to two ways (two values of k) of drawing tangents to the parabola.

---

3. *Obtain the range of x for which the simultaneous conditions $2x^2 - 3x - 3 > 0$ and $|2 - x| > 1$ are satisfied.*

Solution:
We will analyse these inequalities graphically. Consider first the curve $y = 2x^2 - 3x - 3$; it has roots at $x = (3\pm\sqrt{33})/4$ and since it is a parabola with a minimum, the condition $2x^2 - 3x - 3 > 0$ implies the values $x > (3 + \sqrt{33})/4 \approx 2.2$ or $x < (3 - \sqrt{33})/4 \approx -0.68$.

The second condition $|2-x| > 1$ translates to the values, $2-x > 1 \Rightarrow x < 1$ or $2 - x < -1 \Rightarrow x > 3$. The two conditions are

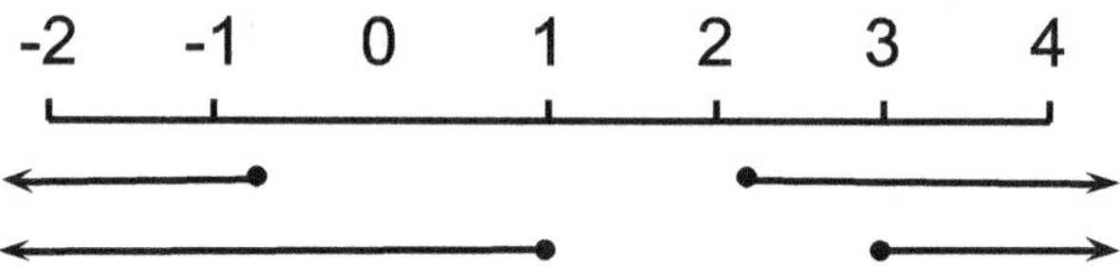

Figure 3.1: Figure for worked example 3.

simultaneously satisfied when $x > 3$ or $x < (3 - \sqrt{33})/4$, see Fig.(4.1).

Answer: $x > 3$ or $x < (3 - \sqrt{33})/4$.

---

4. *By using an appropriate plot, solve $f(x) = 2e^{-x} - x - 1 = 0$ approximately for $x > 0$ (attempt to get one decimal place accuracy).*

Solution:

This equation is difficult, if not impossible, to solve analytically, but an approximate solution may easily be found. First note that since $f(0) = 1$ and $f(1) < 0$, we know that there is at least one zero of $f(x)$ between $x = 0$ and $x = 1$. Then you may use the approximation methods discussed in Chapter(9): See question (12) of Sect.(9.1) or question (11) of Sect.(9.2).

Another way is to use graphical methods. Re-write $2e^{-x} - x - 1 = 0$ as $2e^{-x} = x + 1$ which may be interpreted as the intersection of the two curves $y = 2e^{-x}$ and $y = x+1$. By plotting the two curves on the same graph you can estimate the location of the single intersection point, $x \approx 0.4$. It is useful to check the accuracy of the approximate solution by direct substitution: $f(0.4) \approx -0.06$. (The exact solution is $x = 0.3748...$).

Answer: $x \approx 0.4$.

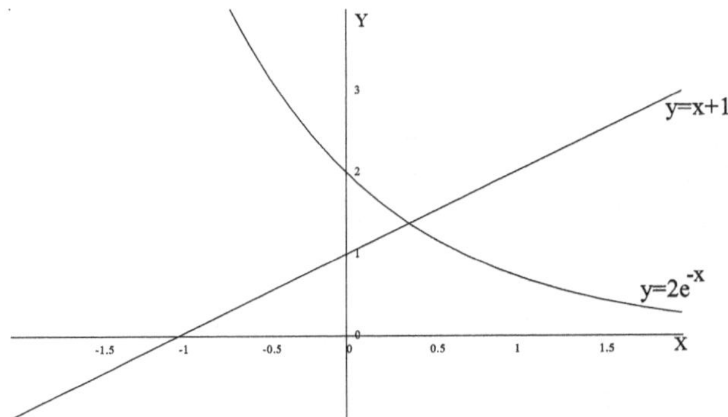

Figure 3.2: Figure for worked example 4.

---

5. *Three toll bridges A, B, C connect an island to the mainland. The vehicles that need to pay toll when crossing a bridge are classified as cars, buses or motorcycles. The first table below represents the number of vehicles of each type that cross each bridge on one particular day. The second table shows the toll charges per vehicle for weekdays and weekends.*

   (a) Construct a 3×3 matrix V, with rows representing different bridges, to represent the information in the first table.

   (b) Construct a 3×2 matrix T to represent the information in the second table.

   (c) Explain what information is contained in the matrix product $P = VT$.

   (d) Use the matrix P to deduce the total toll collected by each bridge if that day was a weekend.

   (e) Use the matrix P to determine the bridge with the highest toll collection if that day was a weekday.

Number of vehicles using each bridge:

	Cars	Buses	Mcycles
A	30	15	20
B	20	10	40
C	15	30	50

Toll charges:

	Cars	Buses	Mcycles
Weekday toll	5	15	2
Weekend toll	3	10	1

Solution:

(a)

$$V = \begin{pmatrix} 30 & 15 & 20 \\ 20 & 10 & 40 \\ 15 & 30 & 50 \end{pmatrix} \tag{3.16}$$

(b)

$$T = \begin{pmatrix} 5 & 3 \\ 15 & 10 \\ 2 & 1 \end{pmatrix} \tag{3.17}$$

(c)

$$P = VT = \begin{pmatrix} 415 & 260 \\ 330 & 200 \\ 625 & 395 \end{pmatrix} \tag{3.18}$$

The left column of P was obtained by multiplying the rows of V by the first column of the toll matrix and it clearly represents the amount of toll collected by each bridge on a weekday. Likewise the second column of P represents the toll collected by each bridge on a weekend (assuming the number of vehicles is the same as indicated in the V table).

(d) A: \$260, B: \$200, C: \$395.

(e) Bridge C.

Answers: See above.

3.3   Exercises

1. Two boats of tourists are next to each other. The passengers of A say to the passengers of B: " If one of you comes over to our boat, then we will have twice as many passengers as you". However the passengers of B reply with a better proposal to A: "If one of you comes over to our boat then each boat will have the same number of passengers. How many passengers are in each boat?

2. The 200 students of a school are teamed up into groups of either 3 or 5. If the total number of teams formed is less than 50, determine a constraint on the number of teams with 3 members.

3. (a) The sum of two numbers is 10, while the sum of their reciprocals is 2. What are the numbers?
   (b) Optional: Sketch the two conditions on a single graph.

4. The sum of two numbers is 18 while the sum of their reciprocals is larger than 4. What is the constraint on the smaller number?

5. Using the equation $y = ax - bx^2$ for the trajectory of ball in Sect.(1), explain how you would analyse Eona's problem mathematically.

6. Find the determinant of each of the matrices A below and find the inverse matrix if it exists; then verify the relation $AA^{-1} = A^{-1}A = I$.

   (a) $\begin{pmatrix} 1 & 2 \\ 3 & 4 \end{pmatrix}$.

   (c) $\begin{pmatrix} -1 & 1 \\ -2 & -2 \end{pmatrix}$.

   (b) $\begin{pmatrix} 1 & 2 \\ 2 & 4 \end{pmatrix}$.

   (d) $\begin{pmatrix} -4 & 3 \\ -2 & 1 \end{pmatrix}$.

7. Find the range of values of x which satisfy the following:
   (a) $-3x + 3 > 2 - x$.

   (b) $x - 1 > 3x - 4 > -2 - x$.
   (c) $5 - 2x > 2 - x > 3x - 4$.

8. Solve the following sets of equations for x, y:
   (a) $2y - 5x = 1$ and $2x - 5y = 3$.
   (b) $2y - 5x = 1$ and $y = 3x^2 - 2$.
   (c) $2y - 5\sqrt{x} = 1$ and $2x - 5y = 3$.
   (d) $\dfrac{1}{y - \sqrt{x}} = \dfrac{1}{\sqrt{x} + 2y} = -1$.

3.4   Problems

1. The 200 students of a school are to be teamed up into groups of either 3 or 5. Let the total number of teams formed be N. Find the smallest value of N for which the number of teams with 3 members exceeds the number of teams with 5 members.

2. A finance company manager decides to reward his 170 employees with a trip to the beach. He also decides to support two of his clients, which are transportation providers, by hiring buses from each of them. The buses from provider X can hold 40 passengers each while those from provider Y can hold 30 passengers each. Determine the minimum number of buses he should hire from each provider if the total number of buses should not exceed 5 (he must hire at least one bus from each provider).

3. In the previous question, suppose there is no restriction on the total number of buses; rather, suppose the manager wishes to spend the same amount of money on both providers. If each bus from provider X costs five times as much to rent as a bus from provider Y, what is the minimum number of buses he should hire from each provider?

4. The sum of the squares of two positive numbers is 100, while the sum of the squares of their reciprocals is 1. What are the numbers?

5. Solve the following set of equations for x, y:
   (a) $y^2 = 2xy + 3x + 2$ and $y = |x - 1|$.

(b) $x^2 - 2y^2 = 3$ and $y = 9 - 2x^2$ and $x < 0$, $y < 0$.
(c) $x + y = 2xy + 3$ and $xy = 4$.
(d) $y = 2x^2 - 3x + 3$ and $y = |3 - x|$.

6. Find all values of p for which the following simultaneous equations have only one solution (x, y) for a fixed p: $2y - px = 1$ and $y = 3x^2 + px + 1$.

7. Find the range of values of x which satisfy the following:
(a) $2x^2 - 3x - 4 < 0$ and $2 - x < 0$.
(b) $x^2 - x - 1 > 4$ and $3x^2 - 4 > x - 2$.
(c) $2x^2 - 3x + 3 > 2 - x > 3x - 4$.

8. A two digit number has the following properties: (i) The number is four times the sum of its digits, and (ii) Five times the number equals twelve times the sum of the square of the digits of the original number. What is the number?

9. A rectangle of unit area has a perimeter of length P. If $P^2 = 24$, determine the dimensions of the rectangle.

10. Five participants, labelled a, b, c, d, e make it to the finals of a singing contest. Each participant is scored by two judges, J and K, on three criteria: style S, vocals V, and originality G. The tables below show the scores given by the judges.
(a) Construct two 3×5 matrices, J and K, to represent the scores of the two judges.
(b) The total raw score for each criteria is the sum of each judges' score for that criteria. Explain how you would construct a matrix T from J and K to represent the total raw score for each criteria for the participants. Determine the matrix T explicitly.
(c) The final score for each participant is a weighted average of their total raw scores for each criteria: The style factor is multiplied by 0.2, the vocal factor by 0.5 and the originality factor by 0.3. Construct a 1×3 matrix to represent the "weight" matrix W which when multiplied to T would give the final score F.
(d) Determine the matrix F explicitly.
(e) The winner is the participant with the highest final score.

Who was the winner?

	a	b	c	d	e
S	5	1	3	2	4
V	4	3	5	2	5
G	3	2	5	3	4

Scores by Judge J.

	a	b	c	d	e
S	4	2	3	3	3
V	2	3	4	3	5
G	4	3	4	2	4

Scores by Judge K.

11. (a) Students at a fund-raising fair sell three home-made drinks named, "Cool", "Wow" and "Zing" using the same natural ingredients but in different proportions as listed in the table below; construct a 3×3 matrix to represent that information.

(b) All the ingredients have to be bought from one of two suppliers (A, B) whose prices (per glass of drink) are listed in the second table. Calculate, using appropriate matrix multiplication, how much each glass of drink would cost to make if it was made using materials wholly provided by one or the other supplier (assume water is provided free by the school and no labour or other costs are involved.)

(c) The drinks made using ingredients from supplier A are sold at one stall, the "Apex", while the drinks made using ingredients from supplier B are sold at another stall, "Bravo". Each drink is sold at double its cost price and the number of glasses sold at each stall is listed in the table below. Calculate, with explanation, the total profit from the sales at (i) stall Apex , (ii) stall Bravo.

Ingredients table (in dimensionless units):

	Cool	Wow	Zing
Colouring	1	2	3
Flavouring	2	3	1
Sweetener	3	1	2

Ingredients cost (in dollars) per unit ingredient:

	Colouring	Flavoring	Sweetener
Supplier A	0.1	0.2	0.3
Supplier B	0.15	0.15	0.25

Sales table (number of glasses):

	Cool	Wow	Zing
Apex	5	10	20
Bravo	10	15	15

3.5 Challenges

1. Find all non-negative integer solutions x, y of the following equation: $3x + 5y = 66$.

2. A kindergarten teacher spent exactly \$13 to buy packaged drinks of three different flavours for her 30 students: Grape, orange and strawberry. Each packet of grape drink cost her \$1, while for the same price, \$1, she could get either two packets of orange drink or three packets of strawberry drink. The drink packets of each flavour were distributed equally to each student. What did each student get?

3.6 Inquire and Investigate

1. Are there efficient ways of solving N linear equations in N unknowns for $N \geq 3$?

2. In general, if there are more unknowns than the equations that determine them, one may expect there to be more than one solution unless there is a restriction, for example to the set of of positive integers. Such equations are known as Diophantine equations. An example is question (C2). Another example is to find all right-angled triangles with integral sides: Can you do it?

3. Study a proof by Fermat of his Last Theorem for the case $n = 4$.

Breathe
A simple and effective technique for dealing with anxiety or tension is to breathe deeply and slowly, with minimal movement of the shoulders or chest, focusing on the abdomen. Keep your eyes closed when you do so. Repeat for a few rounds.

Chapter 4

Trigonometry

Eureka! Eureka!
Archimedes

4.1   Introduction

Suppose that you wanted to measure the height of a tall tree. Could you do it by staying on the ground and without using any electronic equipment? Trigonometry is the mathematical tool that you would need (see exercises). Indeed, trigonometry was invented to solve some similar problems in geography and astronomy.

Given a right-angled triangle labelled by its vertices (and thus angles) ABC, with $C = 90°$, the trigonometric functions $\sin A$ ("**sine A**") and $\cos A$ ("**cosine A**") are initially defined by the ratios $\sin A = a/c$ and $\cos A = b/c$, where the small case letters denote lengths opposite the corresponding angles, see Fig.(4.1). (Using properties of similar triangles you can verify that the values of the trigonometric functions do not depend on the size of the triangle.)

From the above triangle one deduces the identity $\sin(90° - A) = \cos A$ and by using Pythagoras theorem you can verify another identity: $\sin^2 A + \cos^2 A = 1$.

Although we used a particular triangle to introduce the trigonometric functions, those functions may be defined more generally, including angles larger than $90°$. For the generalisation, consider a circle with

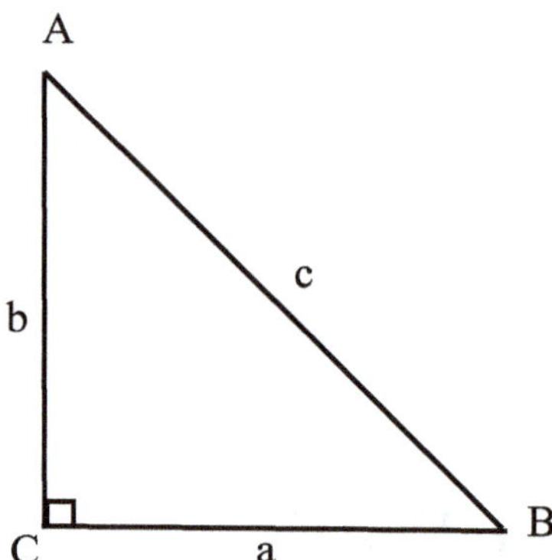

Figure 4.1: A right-angled triangle.

centre at O and radius R as shown in Fig.(4.2). Let P be a point on the circumference with coordinates (x, y) and denote by θ the angle that the line OP makes with the positive $x-$axis.

By **convention**, angles are positive when one moves counter clockwise away from the positive x-axis, and vice versa. Define $\sin\theta = y/R$ and $\cos\theta = x/R$. Note that $\sin\theta$ is positive in the first quadrant $(0 \le \theta \le \pi/2)$ and second quadrant $(\pi/2 \le \theta \le \pi)$ while $\cos\theta$ is positive in the first and fourth quadrant $(3\pi/2 \le \theta \le 2\pi)$. So, $\sin(-\theta) = -\sin(\theta)$, $\cos(-\theta) = \cos(\theta)$ and both functions range over the interval $[-1, 1]$.

With the above definition of the trigonometric functions for all angles, we see that they are **periodic**, with period $360°$, that is $\sin(\theta + 360°) = \sin(\theta)$ and $\cos(\theta + 360°) = \cos(\theta)$. It is this periodic nature of trigonometric functions that makes them very useful in the description of numerous oscillatory phenomena in Nature, science and engineering, such as the motion of tides, waves, and alternating currents. Phenomena that are described by an equation of the form $y(x) = A + B\sin(kx + C)$ are called **sinusoidal**.

The other common trigonometric function, $\tan A$ ("**tangent A**") is defined as the ratio $\dfrac{\sin A}{\cos A}$; it has a period of $180°$ and ranges over the whole real line.

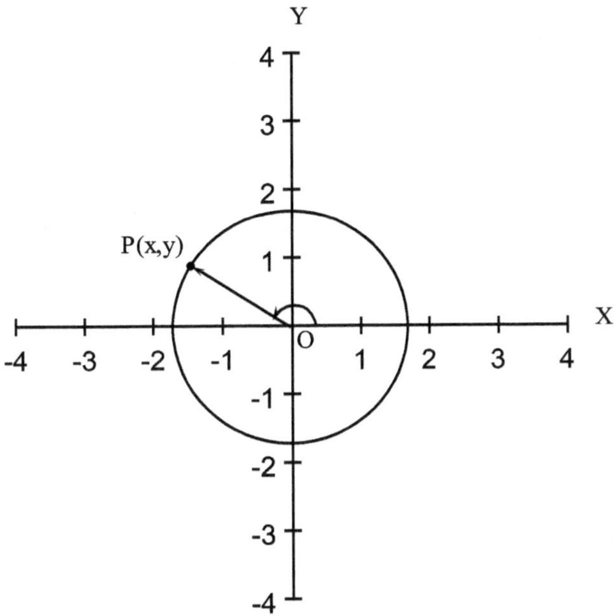

Figure 4.2: A circle is used to define the trigonometric functions for all angles.

The notation for the **inverse** sine function is $y = \sin^{-1} x$ which refers to values of y for which $\sin y = x$. Since the trigonometric functions are periodic, the inverse functions are multi-valued and so usually one restricts the inverse functions to their **principal values** (so that there is a unique inverse): For $y = \cos^{-1} x$ the values lie in the range $0 \le y \le \pi$ while for $\sin^{-1} x$ and $\tan^{-1} x$ functions the range is $-\pi/2 \le y \le \pi/2$.

The functions $\csc A$ (also written as cosec A), $\sec A$ and $\cot A$ are reciprocals of the $\sin A$, $\cos A$ and $\tan A$ functions. (Do not confuse these reciprocals with the inverse functions of the previous paragraph).

Finally, we note that for angle measurements, the dimensionless unit **radian** is useful. Recall that in radians, $\theta = s/R$ where s is the arc length of circle subtended by that angle. Therefore 2π radians equals $360°$.

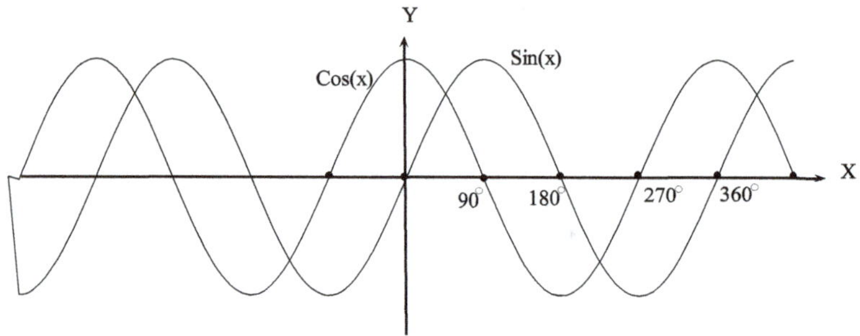

Figure 4.3: The sin and cosine functions

> **Archimedes**
> Archimedes was an extremely creative and influential figure who
> lived in Greece around 250 B.C.E.. He contributed enormously to
> mathematics, science and engineering: The Archimedes principle,
> the Archimedes screw, the lever principle, and the Archimedean
> spiral are some examples of his work. In mathematics, he used
> ingenious techniques, some of which were similar to those of later
> integral calculus, to calculate the areas and volumes of geometrical
> figures. He proved that the area of a circle equalled πr^2 and gave
> accurate estimates for the value of π: $3\frac{1}{7} < \pi < 3\frac{10}{71}$.

4.2   Relations and Properties

At first sight, there might appear to be a large number of trigonometric
identities, but as hinted in the last section, only a few are basic, for
example

$$\sin^2 A + \cos^2 A = 1. \tag{4.1}$$

$$\sin(-A) = -\sin(A). \tag{4.2}$$

$$\cos(-A) = \cos A. \tag{4.3}$$

$$\sin(A + B) = \sin A \, \cos B + \cos A \, \sin B. \tag{4.4}$$

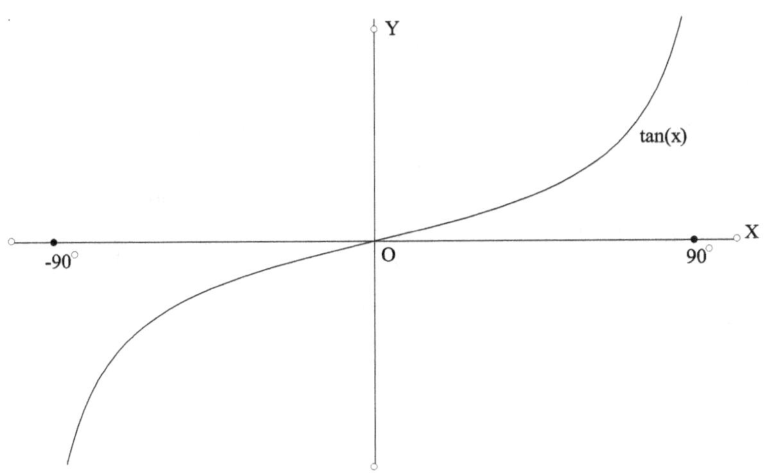

Figure 4.4: The "tan" function

While the first three identities follow from their definition via the unit circle, you may proof eq.(4.4) by constructing a right-angled triangle with $(A+B)$ as one angle and analysing divisions of that triangle. An alternate proof of (4.4) is given in the next sub-section.

Using the above relations one may derive countless others, the more common of which are listed in problems such as (P2) and (P10); we encourage you to try some of those problems to get a feel for how the different trigonometric functions are related to each other. For convenience, the various identities are summarised in one of the appendices. (In actual problem solving, you may quote many of the standard identities without proof.)

Finally, the periodicity and symmetry of the trigonometric functions means that their values for angles larger than $90°$ are related to their values for an angle smaller than $90°$, for example, using the unit circle definition, $\sin 120° = \sin 60°$. The last may also be established algebraically using the addition formula in P(2): $\sin 120° = \sin(180° - 60°) = \sin 60°$; alternatively we could also have used $\sin A = \cos(90° - A)$ to get $\sin 120° = \cos(-30°) = \cos 30° = \sin 60°$.

4.2.1   Properties of Triangles

Bear in mind that the relations in the previous sub-section are properties of the trigonometric functions, holding for all angles.

One may also derive trigonometric relations which hold specifically for triangles. For a right-angled triangle, which is quite special, a popular mnemonic is **"soh-cah-toa"**, which summarises the rules

$$\sin = \frac{Opposite\ side}{Hypotenuse}, \quad \cos = \frac{Adjacent\ side}{Hypotenuse} \quad \text{and} \quad \tan = \frac{Opposite\ side}{Adjacent\ side}.$$

Other trigonometric relations for triangles usually involve both angles and lengths. Consider the triangle ABC in Fig.(4.5) with the altitude h. Its area is $\frac{1}{2}ah = \frac{1}{2}ac\sin B$, the final expression involving two sides and the included angle.

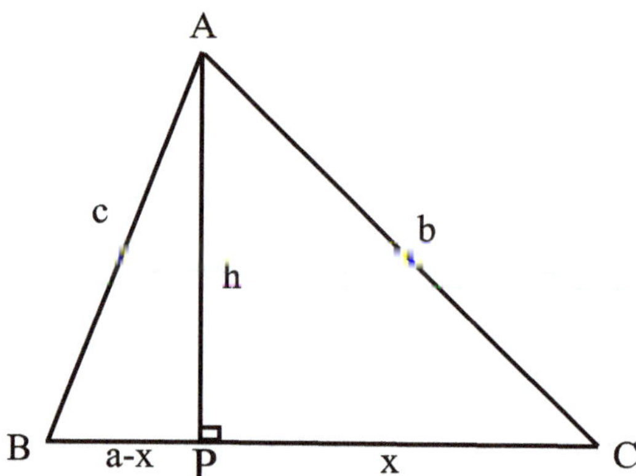

Figure 4.5: The triangle used to derive the sine and cosine rules

Since there is nothing special about those two sides, the same result (area) must hold using any two sides, so it must be that[1]

$$A \;=\; \frac{1}{2}ab\,\sin C = \frac{1}{2}ac\,\sin B = \frac{1}{2}bc\,\sin A. \tag{4.5}$$

Now, dividing eq.(4.5) by $abc/2$ and inverting gives the **sine rule**

---

[1] We have drawn a triangle with acute angles. You can verify that the identities obtained in this section hold also for obtuse angles.

for triangles:

$$\frac{c}{\sin C} = \frac{b}{\sin B} = \frac{a}{\sin A}. \tag{4.6}$$

It turns out that the ratios in (4.6) have a nice interpretation: They equal the diameter of a circle which circumscribes the triangle, the centre of the circle being at the intersection of the perpendicular bisectors of the three sides, see question (C3).

Returning to the triangle, let the altitude h meet side BC at point P with $CP = x$ and $PB = a - x$. Then

$$\begin{aligned} c^2 &= h^2 + (a - x)^2 \\ &= b^2 - x^2 + (a^2 - 2ax + x^2) \\ &= a^2 + b^2 - 2ab\,\cos C, \end{aligned} \tag{4.7}$$

which is the **cosine rule** for triangles. It may be viewed as generalising the Pythagoras theorem to non-right-angled triangles. Of course you can permute the labels of the vertices and find analogous expressions involving $\cos A$ or $\cos B$.

Using the cosine rule and formula (4.5), you may derive Heron's formula (question C2) for the area of a triangle.

Finally, let us prove the identity (4.4) for the sine of a sum of angles. Consider two angles α and β. Construct a triangle ABC as above with $\angle BAP = \alpha$, $\angle CAP = \beta$ and $A = \alpha + \beta$. The area of the triangle may be computed in two ways and equated:

$$\begin{aligned} \frac{1}{2}bc\,\sin A &= \frac{1}{2}ch\,\sin\alpha + \frac{1}{2}bh\,\sin\beta \\ \frac{1}{2}bc\,\sin(\alpha + \beta) &= \frac{1}{2}ch\,\sin\alpha + \frac{1}{2}bh\,\sin\beta \\ \Rightarrow \sin(\alpha + \beta) &= \frac{h}{b}\sin\alpha + \frac{h}{c}\sin\beta \\ \sin(\alpha + \beta) &= \sin\alpha\,\cos\beta + \cos\alpha\,\sin\beta, \end{aligned} \tag{4.8}$$

which is the addition formula (4.4).

4.2.2 Elevation, Depression and Bearing

Trigonometry is useful in surveying and navigation. We introduce some common terminology here.

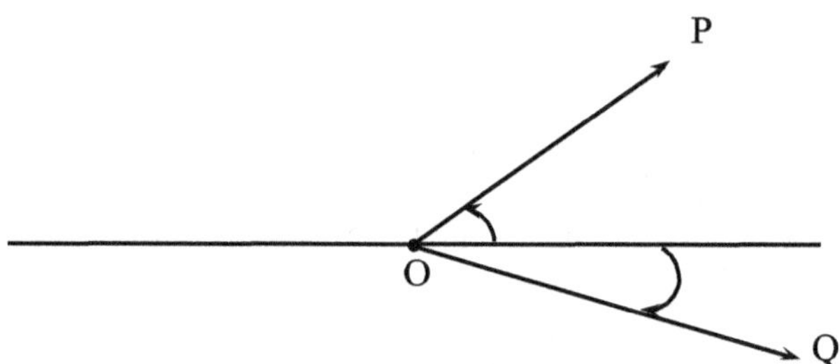

Figure 4.6: Elevation and depression angles relative to the horizontal

If a point P is above the horizontal through the observation point O, then the angle that OP makes with the horizontal is the **angle of elevation** of P. Similarly if Q were below the horizontal then the corresponding angle would be termed **angle of depression**.

A **bearing** denotes a direction relative to North. It is usually expressed by a clockwise angle measured in degrees. For example, 065° would refer to the direction 65° clockwise from North. Such bearings are called "absolute bearings".

Figure 4.7: Absolute bearings are measured clockwise relative to North

It is also convenient to use "relative bearings", which define an angle relative to a chosen axis, for example the axis along which an aircraft is pointing.

4.3 Worked Examples

1. *If $\sin A = 0.3$, find $\cos A$ and $\tan A$ by different methods and compare the answers. If there is more than one value for $\cos A$ or $\tan A$, explain why.*

   Solution:
   Using eq.(4.1), which is the same as using the right-angled triangle of Fig.(4.1), we get $\cos A = \pm\sqrt{1 - (0.3)^2} = \pm 0.9539$. The reason for two values for $\cos A$ is that A could be in the first quadrant, $A = 17.46°$, or second quadrant, $A = 180° - 17.46°$. Then $\tan A = \sin A / \cos A = \pm 0.31$.
   Notice that we did not need to find A explicitly to get the values of $\cos A$ or $\tan A$. The second method is to first find A as $\sin^{-1}(0.3)$ which gives the principal value in the first quadrant, and remember that A could also be in the second quadrant so then $A = \pi - \sin^{-1}(0.3)$.

   Answer: $\cos A = \pm 0.95$, $\tan A = \pm 0.31$.

---

2. *Starting from the identity (4.4) and the other relations in Sect.(2), prove the following identities:*
   (a) $\sin(A - B) = \sin A \cos B - \cos A \sin B$.
   (b) $\sin(90° - A) = \cos A$.
   (c) $\cos(90° - B) = \sin B$.
   (d) $\cos(C + D) = \cos C \cos D - \sin C \sin D$.

   Solution:
   Let us write the starting identity as $\sin(\alpha + \beta) = \sin\alpha\cos\beta + \cos\alpha\sin\beta$. It is valid for all angles. We now sketch the key steps to the proofs, leaving it to you to fill in the gaps:

   (a) Set $\alpha = A$, $\beta = -B$ and use $\sin(-B) = -\sin(B)$, $\cos(-B) = \cos B$.

(b) Set $\alpha = 90°$, $\beta = -A$ and use $\sin(90°) = 1$, $\cos(90°) = 0$.

(c) Set $A = 90° - B$ in the result of part (b).

(d) Set Set $A = C + D$ in the result of part (b), write $(90° - C - D) = (90° - C) - D$ and use part (a), (b) and (c).

(Note that the angles are general. We have used different letters in different identities just to simplify the discussion and avoid confusion.)

---

3. *Re-arrange the following sequence in order, from the largest to the smallest, without using a calculator. Justify your answer:*
$\tan(65°)$, $\sin(25°)$, $\cos(100°)$, $\sin(125°)$.

Solution:
The first step is to relate the expressions with angles larger than $90°$ to values for angles smaller than $90°$. Drawing the unit circle and using the definitions (or addition formulae), we have $\cos 100° = -\cos 80°$, which is $-\sin 10°$ and so negative. Similarly you can show that $\sin 125° = \sin 65°$.

Since the sin function is positive and increasing in the range $0 < \theta < 90°$, so $\sin 125° = \sin 65° > \sin 25° > 0 > -\sin 10° = \cos 100°$.

What about $\tan 65°$? It is $\sin 65°/\cos 65°$, and since $1 > \cos 65° > 0$, so $\tan 65° > \sin 65° > 0$.

Answer: $\tan 65° > \sin 125° > \sin 25° > \cos 100°$.

---

4. *A student observing the angular displacement, θ, of a swinging pendulum finds that it starts from its maximum displacement of $1/6$ radians (from the vertical) at time $t = 0$ and takes 2 seconds*

to complete one oscillation. Assuming the motion is described by the equation $\theta = A\sin(\omega t + \phi)$, with $0 < \phi < \pi$,

(a) Determine the constants A, ω, ϕ.

(b) Is it possible to write the equation of motion in a simpler form?

Solution:

(a) As the maximum value of $|\sin x|$ as x varies is 1, therefore $1/6 = A \cdot 1 \Rightarrow A = 1/6$. Now, at $t = 0$, $\theta = A\sin(\phi)$ and since we are told that θ is maximum at that time, we require $\sin(\phi) = 1$ which implies $\phi = \pi/2$. The equation of motion takes the form

$$\theta = \frac{1}{6}\sin(\omega t + \pi/2).$$

As sin is a periodic function of its arguments, the angular displacement will be periodic. Let T be the time taken for the motion to complete one cycle. From the relation $\sin(x + 2\pi) = \sin(x)$ we see that the difference in the arguments at $t = T$ and $t = 0$ must be 2π: $(\omega T + \phi) - (\omega \cdot 0 + \phi) = 2\pi$, which gives $\omega = 2\pi/T = 2\pi/2 = \pi$.

(b) The equation of motion is $\theta = \dfrac{1}{6}\sin(\pi t + \pi/2)$ but since $\sin(x + \pi/2) = \sin x \cdot \cos \pi/2 + \cos x \cdot \sin \pi/2 = \cos x$, it simplifies to $\theta = \dfrac{1}{6}\cos(\pi t)$.

Answer: (a) $\theta = \dfrac{1}{6}\sin(\pi t + \pi/2)$, (b) $\theta = \dfrac{1}{6}\cos(\pi t)$.

---

5. *Determine the maximum and minimum value of $S(\theta) = 3 + \sin\theta + 4\cos\theta$ in the range $0 \le \theta \le 2\pi$.*

Solution:

The trigonometric functions vary from -1 to $+1$ in that range, but since they do not reach the extremal values at the same time, we use the "R-formula", see problem (P10), to re-write the sum,

$$\sin\theta + 4\cos\theta \equiv R\sin(\theta+\beta)$$
$$= R\sin\theta\cos\beta + R\cos\theta\sin\beta.$$

Comparing the two sides gives $R\sin\beta = 4$ and $R\cos\beta = 1$, and their ratio implies $\tan\beta = 4$, while the sum of squares, $R^2\sin^2\beta + R^2\cos^2\beta$ is $R^2 = 17$.

Therefore the expression is $S(\theta) = 3 + \sqrt{17}\sin(\theta+\beta)$ which reaches a maximum of $3+\sqrt{17}$ and minimum of $3-\sqrt{17}$. Notice that the result does not depend on β. (Try re-solving this problem using calculus).

Answer: Maximum $3 + \sqrt{17}$, minimum $3 - \sqrt{17}$.

---

6. *Solve* $\cos 2x - \cos 3x - \cos x = 0$ *for* $0 \le x \le \pi$.

Solution:
We will use either of the last two of identities in problem (P6) to combine two of the terms. Some thought shows that the terms with odd multiples of x should be combined first, as the result would then be an integral multiple of x:
$\cos 3x + \cos x = 2\cos 2x\cos x$. Therefore

$$0 = \cos 2x - \cos 3x - \cos x$$
$$= -2\cos 2x\cos x + \cos 2x$$
$$= \cos 2x(1 - 2\cos x),$$

which implies $\cos 2x = 0$ or $\cos x = 1/2$. The solutions in the stated range are therefore $x = \pi/4$ and $x = \pi/3$.

Answer: $x = \pi/4,\ \pi/3$.

---

7. *Prove the identity* $\dfrac{1 + \cos A + \cos 2A}{\cot A} = \sin A + \sin 2A.$

Solution:
You can start in any direction that looks promising. We will bring the $\cot A$ over to the right and simplify the result:

$$
\begin{aligned}
\cot A \cdot (\sin A + \sin 2A) &= \frac{\cos A}{\sin A}(\sin A + 2\sin A \cos A) \\
&= \cos A + 2\cos^2 A \\
&= \cos A + 1 + \cos 2A,
\end{aligned}
$$

where in the first and last steps we used the double angle formulae for $\sin 2A$ and $\cos 2A$, see problem (P2). Re-arrange and you are done.

Here is an alternative approach which starts directly from the left-hand-side:

$$
\begin{aligned}
\frac{1 + \cos A + \cos 2A}{\cot A} &= \frac{\sin A}{\cos A}(1 + \cos A + 2\cos^2 A - 1) \\
&= \frac{\sin A}{\cos A}(\cos A + 2\cos^2 A) \\
&= \sin A + 2\sin A \cos A, \\
&= \sin A + \sin 2A \ . \tag{4.9}
\end{aligned}
$$

---

8. *If* $\tan A = p,$ *write* $\sin 2A$ *in terms of* $p.$

Solution:
Here is a purely algebraic approach: We will start with the double angle formula for $\sin 2A$ and try to write it in terms of $\tan A$. Firstly,

$$
\begin{aligned}
\sin 2A &= 2\sin A \cos A \\
&= 2\frac{\sin A}{\cos A}\cos^2 A \\
&= 2\tan A \cos^2 A \ .
\end{aligned}
$$

Next we try to re-write $\cos^2 A$ in terms of $\tan A$; this is easy as it follows from $\sin^2 A + \cos^2 A = 1 \Rightarrow \tan^2 A + 1 = \dfrac{1}{\cos^2 A}$. Hence

$$\begin{aligned} \sin 2A &= \frac{2\tan A}{1 + \tan^2 A}. \\ &= \frac{2p}{1 + p^2}. \end{aligned}$$

Alternatively, assuming A is acute, you can use the right-angled triangle, Fig.(4.1), and $\tan A = p$ to express $\sin A$ and $\cos A$ in terms of p and hence get $\sin 2A$. But then you would also have to argue why the result is true for all A, which the above algebraic approach does easily.

Answer: $\dfrac{2p}{1 + p^2}$.

---

9. *In the triangle ABC, P is a point on AC such that $AP = 10$, $AB = BP = PC$ and $\angle ABP = 110°$,*
   (a) Find the lengths AB and BC.
   (b) Find the area of the triangle ABC.
   (c) Find the shortest distance from C to the line AB produced.
   (Express all answers correct to one decimal place.)

Solution:

The first step is to sketch the triangle and indicate the information provided, see Fig.(4.8) below. The second step: Check your sketch and the information indicated.

(a) Now we are ready to begin. Denote each of the equal sides in the isosceles triangle ABP by x which can be determined using the cosine rule or the sine rule. Since each equal angle is $(180 - 110)/2 = 35°)$, using the sine rule gives

$$\begin{aligned} \frac{x}{\sin 35°} &= \frac{10}{\sin 110°} \\ \Rightarrow x &= 6.1\,. \end{aligned}$$

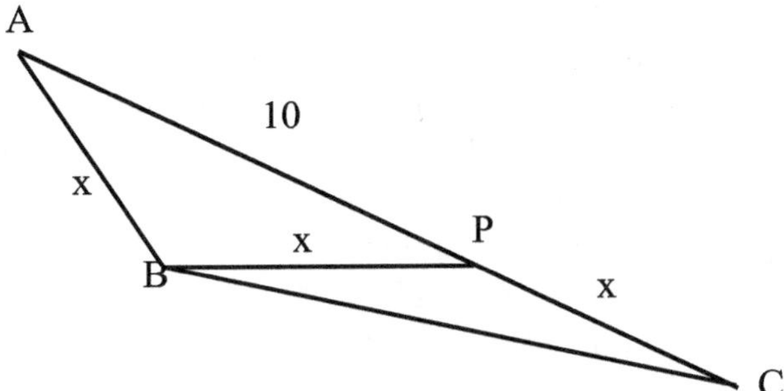

Figure 4.8: Figure for worked example.

You should try re-obtaining x using the cosine rule.

In triangle BPC, $\angle BPC = 180 - 35 = 145°$ and $\angle PBC = \angle PCB = 35/2 = 17.5°$. You may now find length BC using the sine rule: $BC = 11.6$

(b) By (4.5), the area of the triangle is $\frac{1}{2}AB \cdot AC \cdot \sin(\angle BAC)$, which evaluates to 28.2 square units.

(c) The shortest distance from C to the line AB produced will be a line that is perpendicular to AB produced. Denote its length by y. Then the area of the triangle ABC would be $y/2 \times AB$ and this must equal the value found in part (b), giving $y = 9.2$. (Alternatively you may notice from the triangle that the perpendicular distance is simply $AC\sin(\angle BAC)$.)

Answers:$(a)AB = 6.1$, $BC = 11.6$, (b) 28.2, (c) 9.2.

4.4 Exercises

1. Neo decides to measure the height of building. He walks 30 m away from the base of the building, lies down on the ground and uses a home-made device to measure the angle 36° between the

ground and the top of the building.
(a) What is the height of the building?
(b) If, lying at the same place, Neo used his device to look at a point P half-way up the building, what angle would he measure?
(c) If, lying at the same place, Neo used his device to look at a point Q on the building making an angle $18°$ with the ground, how high is Q above ground?
(d) Why are the locations of P and Q different even though $18°$ is half of $36°$?
(e) Optional: Can you make the simple device Neo used? Use it to measure the height of your school building.

2. When light travels from a medium with refractive index n_1 to another with refractive index n_2, such as from air to water, the angle between the light ray and the normal to the interface changes from θ_1 to θ_2. The relationship between the variables is given by Snell's law: $n_1 \sin \theta_1 = n_2 \sin \theta_2$.
(a) Given that the angles are not more than $\pi/2$, show that $\theta_2 > \theta_1$ requires $n_1 > n_2$.
(b) Find the critical incidence angle $\theta_1 \equiv \theta_c$ which occurs when $\theta_2 = \pi/2$.
(c) Optional: What is the significance (or application) of the critical angle?

3. The time-varying coordinates of a particle moving in a constant gravitational field (projectile motion) are given by $x(t) = Ut \cos \theta$ and $y(t) = Ut \sin \theta - \dfrac{gt^2}{2}$ (ignoring air resistance), where U is the initial velocity of the particle, g the acceleration due to gravity, and θ the launch angle relative to the horizontal..
(a) Eliminate the time variable to obtain an equation for the trajectory, $y = y(x)$.
(b) The range R is the horizontal distance between the start of the projectile and the place where it hits the ground. Find an expression for R.
(c) Determine the maximum value of R as θ varies, keeping the other parameters fixed. What value of θ maximises R?

4. Electricity supplied to homes is in the form of "alternating current" (A.C.). The formula for the voltage is $V(t) = V_0 \sin(2\pi f t)$ where $f = 60$ oscillations per second is the frequency, t the time and $V_0 = 240\sqrt{2}$ volts. For $0 \le t \le \dfrac{1}{2f}$,

   (a) Sketch $V(t)$.
   (b) When is the voltage larger than 240 volts?
   (c) When is the voltage less than 120 volts and decreasing?
   (d) Optional: Why is electricity supplied as A.C.?

5. A string is tied at its two ends at $x = 0$ and $x = L$. When plucked, it creates "standing waves" whose vertical displacement is given by $y(x) = A \sin \dfrac{2\pi x}{\lambda}$ where A is the amplitude and λ the wavelength of the wave.

   (a) Using the condition $y(L) = 0$ (since the tied end does not move), find the possible values for λ.
   (b) Optional: How do the words "fundamental" and "harmonics" apply to the above waves?
   (c) Optional: Where else can you find standing waves?

6. (a) Using an equilateral triangle of side 1 (and its bi-section), show that $\sin 30° = \cos 60° = 1/2$ and
   $\sin 60° = \cos 30° = \sqrt{3}/2$.
   (b) Using a right-angled isosceles triangle, show that
   $\sin 45° = \cos 45° = \sqrt{2}/2$.
   (c) Deduce the exact values for $\tan 30°$, $\tan 45°$ and $\tan 60°$.

7. Re-arrange the following sequences in order, from the smallest to the largest, without using a calculator. Justify your answers.
   (a) $\tan(55°)$, $\sin(35°)$, $\cos(35°)$.
   (b) $\sec(15°)$, $\csc(15°)$.

8. Sketch the graphs of the following in the range $0 \le x \le 2\pi$ and determine the maximum and minimum values attained:

   (a) $y = |\sin 2x| - 1$.

   (b) $y = 2 - \cos^2 x$.

   (c) $y = |\sin x + 1| - 1$.

   (d) $y = 2\cos(x/2) - 3$

9. Given $\sin x = 0.2$, and $0 < x < \pi/2$, express exactly the values of
   (a) $\cos x$ and $\tan x$.
   (b) $\sin 2x$, $\cos 2x$ and $\tan 2x$.
   (c) $\sin(x + \pi/3)$, $\cos(2x + \pi/4)$ and $\tan(x + \pi/6)$.

10. Find the smallest positive x which solves each of the following equations:

   (a) $1 + \sin x = 3\cos^2 x$.   (c) $\sin(x° - 20°) = \cos 80°$.
   (b) $\tan^2 x = 2 - 3\sec x$.   (d) $\cos(1 - 2x) = 0.75$.

11. If $f(x) = x\cos x + a\sin x + b\tan x + 5$ and $f(3) = 6$, determine the values of $f(-3)$ and $f(0)$ (a and b are constants).

12. If $\sin^4 x - \cos^4 x = 1/4$, determine the exact values of $\sin x$, $\cos x$ and $\tan x$ for x in the first quadrant.

13. In a triangle labelled by its angles A, B, C, the small case letters a, b, c denote lengths opposite the corresponding angles. Determine all the sides and angles in each case below given the partial information:
   (a) $a = 5$, $b = 7$, $C = 80°$.
   (b) $a = 6$, $b = 7$, $A = 50°$.
   (c) $A = 50°$, $B = 120°$, $c = 7$.
   (d) $A = 50°$, $B = 120°$, $a = 10$.

14. Find the area of each triangle in the previous exercise.

15. (a) A regular pentagon is inscribed in a circle of unit radius. What is the area of the pentagon?
   (b) Another regular pentagon circumscribes the same unit circle. What is the area of the larger pentagon?

16. A circle is inscribed in a regular heptagon of side one. What is the shortest distance from the centre of the circle to the heptagon?

17. From point B, the point C is due east, while A is at a bearing $020°$. Also, $\angle BAC = 50°$. Find the bearings of the following:

(a) A from C.   (c) B from C.

(b) C from A.   (d) B from A.

18. Determine the triangle of largest area that can be inscribed in a given semi-circle of radius R (including the diameter as boundary).

4.5 Problems

1. Eona is on a sailboat, S, some distance from a straight section of the shore. She wishes to estimate how far she is from the shore using angle measurements to two landmarks A and B on the straight section. She measures angle ASP to be $10°$ while BSP is found to be $15°$, P being a point on the shore which makes SP perpendicular to AB. If AB is known to be 500 m long, how far is Eona from the shore?

2. Deduce the following standard identities from the other relations listed in Sect.(2):

   (a) $\sin(A - B) = \sin A \cos B - \cos A \sin B$.

   (b) $\sin(90° \pm A) = \cos A$.

   (c) $\sin(180° \pm A) = \mp \sin A$.

   (d) $\cos(90° \pm A) = \mp \sin A$.

   (e) $\cos(180° \pm A) = -\cos A$ (HINT: Use (b)).

   (f) $\cos(A \pm B) = \cos A \cos B \mp \sin A \sin B$.

   (g) $\sin 2A = 2 \sin A \cos A$.

   (h) $\cos 2A = 2 \cos^2 A - 1 = 1 - 2 \sin^2 A = \cos^2 A - \sin^2 A$.

3. An engineer is asked to design a roller coaster track which has a sinusoidal shape, $y(x) = A + B \cos(kx + C)$ where x is the straight line distance as measured along the ground and y the height of the track above ground. The requirements are: The lowest and highest points of the track should be respectively 20m and 30m above ground; the distance between the peaks of the track should be 50m; the start of the track at $x = 0$ should be 25 above ground.

Can you help the engineer determine the constants A, B, k, C with $C < \pi$?

4. Deduce the following standard identities involving the tangent function from those for sin and cos:

    (a) $\tan(-A) = -\tan(A)$.

    (b) $\tan(A \pm B) = \dfrac{\tan A \pm \tan B}{1 \mp \tan A \tan B}$.

    (c) $1 + \tan^2 A = \sec^2 A$.

    (d) $\tan 2A = \dfrac{2 \tan A}{1 - \tan^2 A}$.

5. Given that $\tan A = 3/7$ and that A is in the first quadrant, express the values of each of the following without using a calculator:

    (a) $\sin A$.

    (b) $\cos(A/2)$.

6. Writing $A = X + Y$ and $B = X - Y$, determine X, Y in terms of A, B. Hence, or otherwise, establish the following standard identities (**Factor Formulae**):

    (a) $\sin A + \sin B = 2 \sin \dfrac{A+B}{2} \cos \dfrac{A-B}{2}$.

    (b) $\sin A - \sin B = 2 \cos \dfrac{A+B}{2} \sin \dfrac{A-B}{2}$.

    (c) $\cos A + \cos B = 2 \cos \dfrac{A+B}{2} \cos \dfrac{A-B}{2}$.

    (d) $\cos A - \cos B = -2 \sin \dfrac{A+B}{2} \sin \dfrac{A-B}{2}$.

7. Given that $\tan A = -3$ and that A is in the second quadrant, express the values of each of the following without using a calculator:

    (a) $\cos 2A$.

    (b) $\sin(A/2)$.

8. When two sound waves of slightly differing frequencies interfere, the result is a wave with periodically varying loudness. The

perceived variation in the loudness is called a "beat". Representing the two waves by the expressions $A_1 = \sin(2\pi f_1 t)$ and $A_2 = \sin(2\pi f_2 t)$, with t the time and $f_2 > f_1$ the frequencies,

(a) Show that the resultant wave, $A_1 + A_2$, may be interpreted as a sound wave of average frequency $(f_1 + f_2)/2$ whose amplitude varies periodically.

(b) If the loudness of sound is proportional to the square of the amplitude, explain why the beat frequency is $f_1 - f_2$.

(c) Optional: How is the phenomena of beats used to tune musical instruments?

9. If A is in the first quadrant and B in the second, with $\sin A = 0.2$ and $\cos B = -0.3$, find the values for each of the following correct to two decimal places:

   (a) $\tan(A + B)$.

   (b) $\sin B + \cos A$.

10. Determine R and α in terms of a, b to establish the following standard identities (**R Formulae**):

    (a) $a\cos\theta \pm b\sin\theta = R\cos(\theta \mp \alpha)$.

    (b) $a\sin\theta \pm b\cos\theta = R\sin(\theta \pm \alpha)$.

    (Answer: $R = \sqrt{a^2 + b^2}$, $\tan\alpha = b/a$.)

11. What are the maximum and minimum values attained by the expressions in problem (10)?

12. Write $\tan 3A$ in terms of $\tan A$.

13. A student comes across a trigonometric identity in an old manuscript: $\sin^2 A - 2\,\mathrm{ver}A + \mathrm{ver}^2 A = 0$.

    (a) Assuming that sin is the usual "sine" function, help the student determine a simple expression for the trigonometric function "ver A" in terms of the common trigonometric functions.

    (b) Optional: Why was the "versine" function once popular?

14. Solve the following equations for $0 \le x \le 2\pi$, by using factorisation or other means:

(a) $\sin(\pi/2 - 2x) + 2\sin(x) = 0.$ (d) $2\sin 2x - \sin^2 x - 3\sin x = 0.$

(b) $\sin(3x/2) - \sin 2x = 0.$ (e) $1 - 2\sin^2 x = 3\cos x.$

(c) $5 - \cot(x) \cdot \cot 2x = 0.$

15. Find the three smallest positive x values for which
$7\cos 3x + 5\sin 2x = 5\sin 3x - 7\cos 2x.$

16. Prove the following identities:

(a) $\dfrac{2}{\tan A + \cot A} = \sin 2A.$

(b) $\dfrac{1 + \tan A}{1 - \tan A} = \sec 2A + \tan 2A.$

(c) $\dfrac{\sec^4 A - \sec^2 A}{\tan^4 A + \tan^2 A} = 1.$

17. Given the function $g(x) = a\sin^2 x - b\cos x + cx$ with $g(2) = 3$ and $g(-2) = 5$,
(a) Determine the value of c.
(b) If also $g(1) = 1$, determine a and b.

18. If $\sin^4 x + \cos^4 x = 2/3$, determine $\sin x$, $\cos x$ and $\tan x$ exactly.

19. Determine the maximum and minimum values of the following for $0 \leq \theta \leq 2\pi$:
(a) $7\sin\theta + 3\cos\theta.$
(b) $\cos\theta - 2\sin\theta.$

20. Sketch the graphs of the following in the range $0 \leq x \leq 2\pi$:

(a) $y = |1 + \sin x + 2\cos x|$.   (c) $y = \sin^2 x + x$.
(b) $y = \sin 2x + \sin 4x$.

21. By using appropriate plots, solve the following equations approximately for the smallest x in the range $0 \leq x \leq 2\pi$:

(a) $\tan x - x = 1$.

(b) $\sin x - \tan 3x + 1 = 0$.

(c) $e^{-x} + \cot x = 0$.

22. A physicist observes standing waves on a plucked string and models their vertical displacement with the equation
$$A(x) = 3\sin\left(\frac{2\pi x}{7}\right) + 5\cos\left(\frac{2\pi x}{7}\right)$$
where $x \geq 0$ is the horizontal coordinate.

(a) What is the maximum vertical displacement of the waves?

(b) What is the smallest x at which there is maximum displacement?

(c) What is the smallest non-zero x at which there is no displacement? (Such points are called nodes).

(d) Optional: Where else can you observe standing waves?

23. A student studies an approximate model describing the height $H(t)$, in metres, of tides along the northern coast of Singland as a function of time t measured in hours,
$$H(t) = 3.5 + 1.5\cos\left(\frac{2\pi t}{24}\right) + 1.5\cos\left(\frac{2\pi t}{12} - 4\right).$$

(a) By writing the sum of two trigonometric functions as a product, show that the heights predicted by the model are bounded. Find those simple bounds. Note: The bounds need not be the extremal heights actually attained.

(b) According to the model, when is earliest time $(t > 0)$ when the tide is at 3.5m?

(c) Optional: What causes tides?

24. In the triangle ABC, $\angle ABC = \pi/2$ and $AB = 10$. The point P on AC such that $AP = 5$. The point Q on BC makes PQ perpendicular to BC and $\angle BAQ = \pi/10$.

(a) Sketch the triangle and indicate all the information provided.

(b) Find AQ, $\angle AQP$ and QP.

(c) The areas of triangles ABC and PQC.

25. In the triangle ABC, $a/b = 2$, $\sin B/\sin C = 0.898$ and $c = 5.57$. Find, to one decimal place,

(a) The value of the smallest angle in the triangle?

(b) The length of the longest side?

(c) The area of the triangle?

26. In the rectangle $ABCD$, P is a point on AD such that $PD = 2$, $BP = 5$ and $\sin(\angle BCP) = 0.2$. Without using a calculator, determine,

(a) The length AB.

(b) The area of the rectangle.

27. (a) The angle bisectors of an equilateral triangle of side 1 meet at C. Find the distances from C to each vertex and each side.

(b) A pyramid is in the shape of a regular tetrahedron: The base and the three faces are identical equilateral triangles of side 1. Determine (i) The height of the pyramid and (ii) The largest angle of elevation from a point on the perimeter of the base to the top of the pyramid.

28. A tourist climbs 100 steps to reach the top of a monument. Each step has a height of 15cm and a depth of 20cm. What is the angle of depression from the top step to the ground in front of the first step, along the line of sight?

29. Eona wishes to determine the height of a telecommunications tower which stands vertically on the roof of a building of height 210 m. She walks 100 m away from the foot of the building and measures the angle of elevation of the top of the telecommunications tower from ground level; it is $65°$.

(a) How tall is the telecommunications tower?

(b) If a cable is stretched from Eona's location to the top of the tower, how long would it be?

30. On the flat roof of Singland's City Hall is a hemispherical dome of radius $20m$. At the highest point of the dome, there is a vertical flag pole whose top, T, is 5m above the dome. A straight cable connects the top of the pole to a point R on the roof.

(a) Explain why the shortest possible cable must touch the dome at a point P.

(b) Find the angle of depression of P from T.

(c) Find the length of the shortest cable possible.

(d) Find the distance of R from the edge of the dome.

31. An equilateral triangle is inscribed in a circle which is itself inscribed in a unit square. What is the area of the triangle?

32. A circle C_1 is inscribed in an equilateral triangle which itself is inscribed in a circle C_2 of area π. What is the area of C_1?

33. Neo wishes to determine the height of building A. He measures the elevation of the roof of A from the base of another building B which is some unknown distance from A; he finds the angle to be $20°$. He then climbs to the top of B which he knows to be 30m high and again measures the angle of elevation of the roof of A: Now the angle is $17°$.

(a) What is the height of building A?

(b) How far is A from B?

34. Three houses A, B and C are located such that B is on a bearing $040°$ from A while C is located due east of A, at a bearing $120°$ from B. A tree T, at a bearing $070°$ from A, is equidistant from houses A and B. If $BC = 50$m, find

(a) AB.

(b) AC.

(c) The distance of the tree from house A.

(d) The distance of the tree from house C.

35. A boy scout walks 1 km north of his campsite to point A, then heads on a bearing $030°$ for 2km to point B, and then turns to bearing $130°$ for another 2km to reach C.

(a) What is the straight line distance from C to his campsite?

(b) On what bearing should the scout head from C to return to camp along a straight path?

36. Denoting $\tan(A/2)$ by t, derive formulae for $\sin A$, $\cos A$ and $\tan A$ in terms of t.

4.6   Challenges

1. In a general triangle with sides a, b, c and corresponding opposite angles A, B, C, prove Mollweide's formulae:

$$\frac{a+b}{c} = \frac{\cos\left(\frac{A-B}{2}\right)}{\sin\left(\frac{C}{2}\right)} \quad \text{and} \tag{4.10}$$

$$\frac{a-b}{c} = \frac{\sin\left(\frac{A-B}{2}\right)}{\cos\left(\frac{C}{2}\right)}. \tag{4.11}$$

Hence, or otherwise, obtain the "tangent rule" for triangles:

$$\frac{a-b}{a+b} = \frac{\tan[\frac{1}{2}(A-B)]}{\tan[\frac{1}{2}(A+B)]}. \tag{4.12}$$

2. (a) Derive Heron's formula, $A = \sqrt{s(s-a)(s-b)(s-c)}$ which gives the area of a triangle in terms of its three sides a, b, c; s is half the perimeter.
(b) Using Heron's formula or otherwise, prove that for a triangle with given perimeter, the area is maximised when the triangle is equilateral.

3. Prove that the perpendicular bisectors of the sides of a triangle meet at one point (called the circumcentre) which is the centre of the circumscribing circle. Find the diameter of the circle in terms of a and A, where a is any side of the triangle and A the corresponding opposite angle.

4. Prove that the angle bisectors of a triangle meet at one point (called the incentre) which is the centre of the inscribed circle. Find the radius of the circle in terms of the perimeter of the triangle and its area.

5. A median in a triangle is a line which starts at one vertex and bisects the opposite side. Prove that the medians of a triangle meet at one point C (called the centroid). Determine the ratio by which C divides each median.

6. Prove Ptolemy's Theorem: For a quadrilateral $ABCD$ inscribed in a circle, $AB \times CD + BC \times DA = AC \times BD$, where AB denotes the length of the straight line from A to B etc.

4.7 Inquire and Investigate

1. Explain in detail how triangulation may be used to measure the distance of stars from Earth.

2. How did Archimedes prove the formula for the area of a circle, and how did he bound the value of π?

3. Investigate "Spherical Trigonometry", the study of the relationship between sides and angles of triangles drawn on the surface of a sphere. How do the relations differ from plane trigonometry? What applications does this field have?

4. Study the use of trigonometry in Fourier Series and the applications of the later.

5. (a) Is it possible to define the sin of an imaginary number, for example, $\sin(2i)$ where $i \equiv \sqrt{-1}$?
   (b) Would the trigonometric identities hold for imaginary angles?
   (c) Would the trigonometric identities hold for complex angles?

Sleep
Adequate sleep, both in quantity and quality, is essential for maintaining positive health. If you have difficulty sleeping, try some gentle stretching to relieve muscular tension, and deep breathing to de-stress, before bedtime.

Chapter 5

Coordinate Geometry

> *It is not enough to have a good mind.*
> *The main thing is to use it well.*
> *Rene Descartes*

5.1 Introduction

Neo looks at the tourist map of a city he is exploring to locate a historical landmark; the guide at the bottom of the map says it is located in grid "C4". Using that map reference, Neo is able to find his way to his destination.

The coordinates "C4" are a pair, the letter referring, for example, to a horizontal label on the map while the number gives the vertical label. More generally, one may use a pair of real numbers (x, y) to denote displacements from two given perpendicular axes and hence locate any point on a plane; such coordinates are called **Cartesian coordinates**.

The use of Cartesian coordinates allows geometrical problems to be studied using the powerful methods of algebra and calculus. For example, by analysing a huge amount of observational data, Kepler concluded that the planets make elliptical, rather than circular, orbits around the Sun. This came as a surprise to most people of his day as the circle was considered the "perfect figure", because of its symmetry, and "heavenly bodies" were expected to behave perfectly.

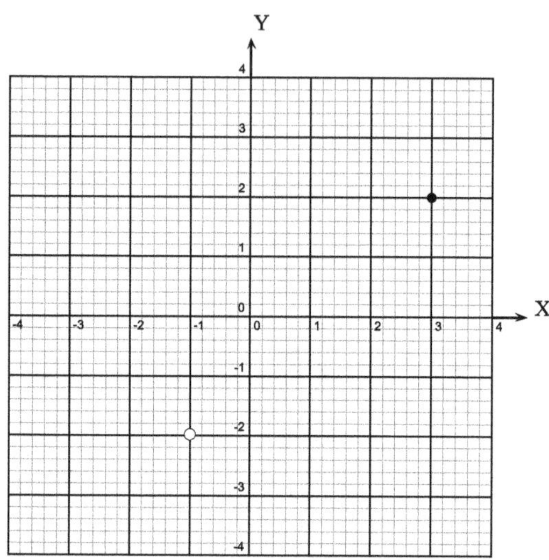

Figure 5.1: The coordinates of the darkened circle are $(3, 2)$ while the white circle is at $(-1, -2)$.

Kepler's results became widely accepted only after Newton provided a mathematical derivation using his newly developed calculus: He showed that an inverse square law of gravity (which may be independently tested) predicted precisely such elliptical orbits.

While "analytical geometry" often makes the analyses of difficult geometrical problems easier, it also works in reverse: Sometimes algebraic problems are easier to understand when they can be visualised through geometry; for example, sketching a curve often helps us deduce its properties with ease.

One important application of coordinate geometry is in the analysis of the relationship between variables. For example, in an experiment one might collect a set of data points (x_i, y_i) for $i = 1, 2, 3....N$, try to deduce a relationship of the form $y = y(x)$ by fitting the data to some theoretical model, and then use the relationship to make some new predictions. In such situations the task is often easier if, through a change of variables from (x, y) to (X, Y), one can get a linear graph: $Y = mX + C$; some of the exercises illustrate this idea.

> **Descartes**
>
> Rene Descartes was a 17th century philosopher and mathematician. He emphasized the importance of logical reasoning for the development of the sciences, and at the same time used mathematical methods to study philosophy. His main contribution to mathematics was the introduction of Cartesian coordinates, a pair of numbers which referred to a point on the plane. His most famous saying, "I think, therefore I am", summarised his insight that, although the senses were unreliable, what he could be sure of was that he was a "thinking thing". Descartes had the habit of getting up late, but when he was in Sweden as a tutor to the queen, who was an early riser, he had to get up at 5 am every day during the winter to teach in a cold palace; unfortunately he caught pneumonia and died from the complication.

5.2 Relations and Properties

We summarise here some relations for figures in a two dimensional plane.

A straight line has constant slope (gradient), so if (x, y) and (x_1, y_1) are any two points on the line, we may write

$$\frac{y - y_1}{x - x_1} = m = \tan \alpha, \tag{5.1}$$

where $0 \le \alpha < \pi$ is the angle the line makes with the positive x-axis. The equation for a straight line may be written in alternate forms, for example as $ax + by + d = 0$, for some constants a, b, d or as

$$y = mx + c \tag{5.2}$$

where m is the slope and c the intercept on the y-axis.

If another line is perpendicular to (5.2), its slope must be $-1/m$ (can you show this?).

The mid-point of a line joining two points (x_1, y_1) and (x_2, y_2) is given by the intuitive formula $\left(\dfrac{x_1 + x_2}{2}, \dfrac{y_1 + y_2}{2} \right)$.

The points on a circle are equidistant from a centre. If (x_0, y_0) are the coordinates of the centre and (x, y) a point on the circle of radius

r, then by Pythagoras' theorem,

$$(x - x_0)^2 + (y - y_0)^2 = r^2 , \tag{5.3}$$

which when expanded may be written in the form

$$x^2 + y^2 - 2ax - 2by + c = 0 , \tag{5.4}$$

for some constants a, b, c. Conversely, given an equation in the form (5.4) you can complete the squares to get $(x-a)^2+(y-b)^2 = a^2+b^2-c$; if $c < a^2+b^2$ then (5.4) represents a circle with centre (a, b) and radius $R = \sqrt{a^2 + b^2 - c}$.

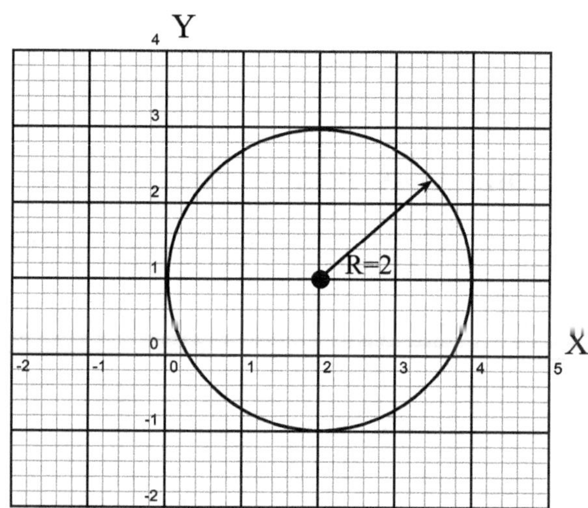

Figure 5.2: A circle with centre at $(2, 1)$ and radius 2.

A tangent to a circle at any point P on its circumference is perpendicular to the radial line OP where O is the centre of the circle. Therefore the **normal** to the circle at P lies along OP.

Given three points $A(x_1, y_1)$, $B(x_2, y_2)$ and $C(x_3, y_3)$, with their relative order being anti-clockwise, the area of the triangle ABC may be obtained from the formula

$$A = \frac{1}{2} |x_1(y_2 - y_3) + x_2(y_3 - y_1) + x_3(y_1 - y_2)| . \tag{5.5}$$

Notice the symmetry in the expression, in particular the cyclic $1 \to 2 \to 3 \to 1$ etc. labeling of indices. The formula is often presented in the "matrix" form

$$A = \frac{1}{2} \begin{vmatrix} x_1 & x_2 & x_3 & x_1 \\ y_1 & y_2 & y_3 & y_1 \end{vmatrix} \tag{5.6}$$

and the terms are generated as follows. First, start at the left of the top row and multiply each term in the top row by a term one step to the right in the bottom row, adding the pieces: $x_1 y_2 + x_2 y_3 + x_3 y_1$. Then start at the right of the top row and multiply each term in the top row by a term one step to the left in the bottom row, adding the pieces: $x_1 y_3 + x_3 y_2 + x_2 y_1$. Finally, subtract the two contributions and include the overall $1/2$ to get

$$A = \frac{1}{2}(x_1 y_2 + x_2 y_3 + x_3 y_1 - x_1 y_3 - x_3 y_2 - x_2 y_1), \tag{5.7}$$

which is the same as (5.5).

The area of a polygon may be determined by dividing the polygon into triangles and using the above formula for each triangle.

5.3 Worked Examples

1. *A curve has equation $x^2 + y^2 - 2x + 4y - 4 = 0$.*
   (a) Show that the curve is a circle and determine its radius R and coordinates of its centre C.
   (b) Determine whether the point P with coordinates $(2, 0.5)$ is inside or outside the circle.
   (c) The line through the origin $O = (0,0)$ and P intercepts the circle at two points A and B with the x coordinate of A smaller than that of B. Determine the coordinates of A and B correct to two decimal places.
   (d) Determine the area of the triangle CAB.
   (e) Determine the shortest distance from C to the line AB produced. (f) Find the equation of the normal through B.
   (g) Find the equation of the tangent at B.

Solution:

(a) In the expression $x^2+y^2-2x+4y-4=0$ complete the square separately for the terms involving x and for the terms involving y to get $(x-1)^2-1+(y+2)^2-4-4=0$ or $(x-1)^2+(y+2)^2=9$ which represents a circle with centre at $C=(1,-2)$ and radius $\sqrt{9}=3$.

(b) Evaluate the distance $CP=\sqrt{(2-1)^2+(0.5+2)^2}=2.69$. It is less than the radius 3 and so P is inside the circle.

(c) The line through O and P has the equation $\dfrac{y-0}{x-0}=\dfrac{0.5-0}{2-0}$ or $y=x/4$. Substitute this into the equation for the circle found in part (a) to solve for the intersection points. We get $(x-1)^2+(x/4+2)^2=9$ or $1.0625x^2-x-4=0$ whose roots are $x=-1.53$ and $x=2.47$. The points A and B are therefore $A(-1.53,-0.38)$ and $B(2.47,0.62)$.

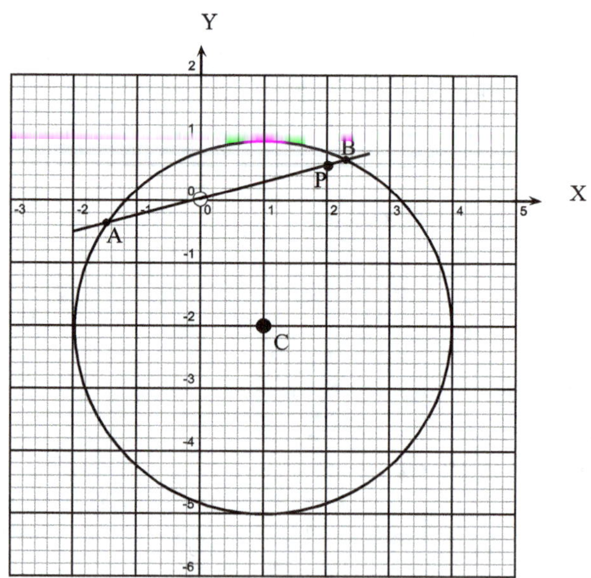

Figure 5.3: Figure for worked example 1.

(d) The vertices of the triangle are

$C(1, -2)$, $A(-1.53, -0.38)$, $B(2.47, 0.62)$. Using (5.5) the area of triangle CAB is found to be 4.50.

(e) The shortest distance from C to AB will be along a straight line that is perpendicular to AB. Denote it by h. Then the area of triangle CAB may be written $h|AB|/2$ and equated with the result of part (d) to obtain h.
Since $|AB| = \sqrt{(2.47 + 1.53)^2 + (0.62 - 0.38)^2} = \sqrt{17}$, therefore $h = 2.19$.

(f) The normal through B is the line that is perpendicular to the tangent at B, and for a circle the normal coincides with the line CB produced (which is the radial direction). The equation of the normal is therefore $\dfrac{y + 2}{x - 1} = \dfrac{0.62 + 2}{2.47 - 1}$ or $y = 1.78x - 3.78$.

(g) Since the slope of the normal at B is 1.78, the slope of the tangent at that point would be $-1/1.78 = -0.56$. The equation of the tangent through B is therefore $\dfrac{y - 0.62}{x - 2.47} = -0.56$ or $y = 2.00 - 0.56x$.

Answers:(a) $C = (1, -2)$, $R = 3$, (b) inside,
(c) $A(-1.53, -0.38)$, $B(2.47, 0.62)$,
(d) 4.50, (e) 2.19, (f) $y = 1.78x - 3.78$, (g) $y = 2.00 - 0.56x$.

---

2. *Two variables u and v in an optics experiment are related by the equation $uv - 2(u + v) = 0$. A student draws a graph of $y \equiv 1/u$ against $x \equiv 1/v$. Show that the expected graph will be that of a straight line and determine its slope.*

Solution:
Dividing the equation through by uv we get $1 - 2\left(\dfrac{1}{v} + \dfrac{1}{u}\right) = 0$

which may be re-arranged to give $y = \dfrac{1}{2} - x$, the equation of a straight line with slope -1.

Answer: $y = \dfrac{1}{2} - x$. Slope is -1.

---

3. *Find the intersection points of the circle $x^2 + y^2 = 9$ with the hyperbola $xy = 1$.*

Solution:

One needs to solve a pair of simultaneous equations. Substituting $y = 1/x$ from second equation into the first gives $x^2 + x^{-2} = 9$. Set $x^2 = u$. Then the equation that is to be solved is $u^2 - 9u + 1 = 0 \Rightarrow u = (9 \pm \sqrt{77})/2 = 8.89$ or 0.11. Hence $x = \pm 2.98$ and $y = \pm 0.34$ or $x = \pm 0.34$ and $y = \pm 2.98$.

Notice the symmetry of the solution set which reflects the symmetry of the original equations. (Sketch the curves and note the intersection points.)

Answer: $(\pm 0.34, \pm 2.98)$ or $(\pm 2.98, \pm 0.34)$.

5.4   Exercises

1. Two points on the plane are $C(3, 5)$ and $D(4, 7)$. The origin is $O(0, 0)$. Find

   (a) The coordinates of the mid-point of CD.

   (b) The equation of a line that is parallel to CD and passes through the origin, O.

   (c) The equation of a line that passes through C and is perpendicular to CD.

   (d) The area of the triangle OCD.

   (e) The shortest distance from the origin to the line CD produced.

2. In each case below, find the equation of a straight line that passes through the point $(2, 3)$ and

(a) is parallel to the line $2x + 3y - 1 = 0$.

(b) is perpendicular to the line $2x + 3y - 1 = 0$.

(c) Sketch a diagram showing all the lines mentioned in this question.

3. The line $2x + 3y - 1 = 0$ intercepts the x axis at P and the y axis at Q. Determine

   (a) The coordinates of P and Q.

   (b) The coordinates of the midpoint, R, of the line PQ.

   (c) The equation of the line through R that is perpendicular to PQ.

   (d) The area of the triangles OPR and OQR where $O = (0,0)$.

   (e) Sketch a diagram showing all the points mentioned in this question.

4. Determine the values of m in each case below so that the line $y = mx - 2$ intercepts the circle $(x - 2)^2 + (y - 1)^2 = 9$ at

   (a) No points.

   (b) At exactly one point.

   (c) At two points.

5. The astronomer Kepler discovered a relationship between the period, T, of a planet's orbit around the Sun and its mean distance, R, from the Sun: $T^2 \propto R^3$. If, given a table of (T, R) values, you had to deduce the proportionality constant, what graph would you plot? Explain.

6. The table below shows the distances of the planets ($a = $ semi-major axis) from the Sun and their orbital period T.

   (a) Plot $\log T$ against $\log a$ and estimate the straight line that gives the best-fit to the data. (Does it matter which base you use for the logarithm?)

   (b) Hence deduce an equation of the form $T = f(a)$.

   (c) How does your result compare with Kepler's law (see previous question) ?

   (d) Optional: What is the physical significance of the proportionality constant in Kepler's law?

Planetary data:

Planet	a	T
Mercury	0.39	0.24
Venus	0.72	0.61
Earth	1.00	1.00
Mars	1.52	1.88
Jupiter	5.20	11.86
Saturn	9.54	29.46
Uranus	19.19	84.01
Neptune	30.06	164.79
Pluto	39.48	248.54

7. For analysing the data of a experiment in the special theory of relativity, a student is asked to plot the variable β^2 against v^2 expecting a straight line graph. If the points $(v, \beta) = (0, 1)$ and $(c, 0)$ lie on the line, with c (representing the speed of light in vacuum) a constant,

   (a) Express β in terms of v if $\beta > 0$.

   (b) Hence deduce the maximum value of v if β is real.

8. The intensity of radiation I that penetrates a shielding material of thickness d is given by the relation $I = Ae^{-\mu d}$ where μ is the "attenuation coefficient".

   (a) Show that a plot of $\ln(I/A)$ versus d is expected to be a straight line with slope $-\mu$.

   (b) The data in the table below was obtained using a X-ray source. Which of the two materials, lead or concrete, is the better shield?

   (c) Deduce the corresponding attenuation constant μ for lead or concrete shielding.

   (d) What thickness of concrete would give the same shielding effect as 1 mm of lead for this x-ray source?

   (e) Optional: Where is shielding from X-rays commonly used?

d (mm)	(I/A) (lead)	(I/A) (concrete)
0.5	0.277	0.98
1.0	0.067	0.96
1.5	0.024	0.93
2.0	0.0048	0.91
2.5	0.0017	0.89
3.0	0.00035	0.87
3.5	0.00014	0.85

Data for E(8).

9. A particle moves in the plane in such a way that both its x and y coordinates are oscillatory: $x = A\sin\omega t$ and $y = A\cos\omega t$, where t is the time variable and A, ω are constants. Show that the trajectory of the particle is a circle and determine the radius and centre of the circle.

10. The y-axis is tangent to a circle C. On the circumference of C, the point with the largest y-coordinate is $P(3, 10)$. Determine, with the help of a sketch,
(a) The equation of the circle.
(b) The point on the circle which has the lowest value for its y-coordinate.
(c) The equation of a circle obtained by reflecting C in the x-axis.
(d) The equation of a circle obtained by reflecting C in the y-axis.

11. A rectangle $ABCD$ has two of its perpendicular sides along lines with equations $2x+3y-1 = 0$ and $3x-2y-2 = 0$, which intersect at point A. The vertices are labelled clockwise and $C = (3, -1)$.
(a) Determine the coordinates of A.
(b) Determine the coordinates of B, and D.
(c) Determine the area of the rectangle.

12. Four points on the plane are $A(1, a)$, $B(b, 2)$, $C(3, 5)$ and $D(4, 7)$ where a, b are constants to be determined in each case below:
(a) The four points lie on a straight line.
(b) AC is perpendicular to CD and AB is perpendicular to AC.
(c) $b = a - 1$ and A, B, C are collinear.
(d) $b = a + 1$ and length $AB =$ length BC.
(e) $a = 2b$ and the area of triangle ABD is 10.

5.5   Problems

1. In analysing the data of a laboratory experiment, a student is asked to plot the variable $\ln y$ against x^2. The data seem to fit a straight line which makes an angle of $30°$ with the horizontal axis. The point $(x, y) = (0, 2)$ lies on the line. Determine the relationship between y and x.

2. When Neo plots the function $y(x)$ against x, he obtains a straight line which passes through the points $(x, y) = (1, 2)$ and $(4, 10)$. The variable y is related to x by $y = z/x - 1$ where z is also a variable.
   (a) Find the explicit relationship between z and x.
   (b) What value does z take when $x = 3$?

3. When Eona plots y against x, she obtains a straight line which passes through the points $(x, y) = (0, 5)$ and $(2, 3)$. The variable x is actually related to another variable z by $x = \sqrt{4 - z^2}$.
   (a) Find the explicit relationship between y and z.
   (b) Show that the curve in part (a) is actually the lower half-circle with centre at $(x, y) = (0, 5)$ and radius 2. Sketch the curve

4. A straight line of slope m passes through the point $(1, -1)$. It also intersects the curve $y = x^2/(1+x)$ at two points. Find all possible values of m which make the x-coordinate of one intersection point positive.

5. The curve C with equation $(y + 1)(x + 3) = 2$ intercepts the line $2x + 3y - 1 = 0$ at two points P and Q. Determine
   (a) The coordinates of the points P and Q.
   (b) The length of the line PQ.
   (c) The area of the triangle OPQ where $O = (0, 0)$ is the origin.
   (d) The shortest distance from O to PQ.

6. Find the intersection points of the circle $x^2 + y^2 = 4$ with the parabola $y = \sqrt{x}$.

7. A circle has equation $x^2 + y^2 + 2x + 4y - 4 = 0$. A square $ABCD$ is inscribed in the circle with $A = (-1, 1)$. Determine

(a) The coordinates of the other vertices of the square.

(b) The area of the square.

8. The point $P = (-1, 1)$ lies on the circumference of a circle $x^2 + y^2 + 2x + 4y - 4 = 0$ with centre at C. The point Q also lies on the circumference and the length $PQ = 4$.

(a) Determine the coordinates of Q if its x-coordinate is positive.

(b) A polygon $PRQC$, labelled by its vertices, has the following lines as boundaries: The normals to the circle at P and Q, and the tangents to the circle at P and Q. Determine the coordinates of vertex R and the area of the polygon.

9. A circle with centre at $C(2, 1)$ has a tangent whose equation is $y = 5 - x$. Find

(a) The radius of the circle.

(b) The point T_1 on the circle where the tangent touches.

10. (a) Continuing the previous problem, three other tangents are drawn on the circle such that the four lines coincide with the sides of a square that circumscribes the circle. The tangents touch the circle at the four points T_1, T_2, T_3, T_4, labelled cyclically.

(a) Find the coordinates of the four points.

(b) Find the area of the square $T_1 T_2 T_3 T_4$.

11. A circle $x^2 + y^2 - 2x - 4y - 4 = 0$ has a tangent with slope 2. Find

(a) The equation of the tangent, given that its y-intercept is positive.

(b) The point, P, of contact of the tangent to the circle.

(c) The equation of the normal through P.

(d) The area of the triangle OPC where O is the origin and C the circle's centre.

(e) The equation of a circle with centre at P which passes through the origin.

5.6   Challenges

1. Given two straight lines on the plane, find the angle between them in terms of their respective slopes m_1 and m_2.

2. Find an expression for the shortest distance from a given point to any given straight line.

3. An ellipse is defined with reference to two focal points, F_1 and F_2: It is the curve whose points P satisfy the distance relation $F_1 P + F_2 P = 2a$, with a fixed. If coordinates are chosen on the plane so that $F_1 = (-f, 0)$ and $F_2 = (f, 0)$, with $f = ea$, $0 < e < 1$ find,
   (a) The coordinates where the ellipse cuts the y-axis.
   (b) Show that the equation of the ellipse may be written as
   $\left(\dfrac{x}{a}\right)^2 + \left(\dfrac{y}{b}\right)^2 = 1$ where b is defined through one of the intercepts.
   (c) Hence, show that the ellipse may be obtained from a circle by stretching one coordinate axis relative to the other.

4. Express the equation for a line and a plane in three dimensional space in terms of Cartesian coordinates.

5.7   Inquire and Investigate

1. The Cartesian coordinate system is an example of an orthogonal coordinate system: The coordinate axis are perpendicular to each other.
   (a) Study other orthogonal coordinate systems in two and three dimensions (such as polar and spherical), including the transformation rules from one system to another.
   (b) Discuss some applications in which non-Cartesian orthogonal coordinates are more useful.

2. Are there realistic situations where non-orthogonal coordinate systems are useful? Study one example in detail.

3. Are there problems that are easy to solve geometrically, but difficult algebraically? What about the reverse situation? Give explicit examples.

4. There are several formulae for the area of a triangle. Study the derivation of some of these formulae and discuss their possible practical applications.

5. The circle is an example of a "conic section".
   (a) Investigate that term and find other curves that are also conic sections.
   (b) How do the equations of the other conic sections differ from that of a circle?
   (c) Do all the conic sections appear in Nature?
   (d) Are there polynomial equations in x and y which are of degree 2 but do not correspond to conic sections?

Nutrition

The body affects the mind and vice versa. Eat more fresh fruits and vegetables, drink sufficient water and reduce your intake of processed foods which contain large amounts of synthetic additives.

Chapter 6

Plane Geometry

Let no one ignorant of geometry enter here.
Plato

6.1 Introduction

Geometry started as a practical concern, to solve such problems as determining the area of a farmer's field, and hence the amount of taxes to levy on it by the state. However, soon the subject became interesting in its own right, leading to the study of various geometrical figures and their properties.

Euclid made a systematic summary of the knowledge existent in his time, using an approach which set the tone for the development of mathematics: "Axioms (postulates), Proof, Theorem". That is, starting from some "self-evident" or defining propositions, one uses valid rules to deduce some new results, called theorems.

But how does one know which new result to aim for? This requires some experimentation and experience, leading to the formation of a conjecture (a guess) which one then tries to prove into a theorem, or dis-proof by finding a counter-example.

Thus one of the benefits of the studying geometry is to gain an appreciation of the methodology of formal mathematics. In fact by pondering over the fundamentals of Euclidean geometry, mathematicians realised that they could construct other abstract forms of geometry where the "parallel lines postulate" need not hold. Such geometries turned out to be precisely what Einstein needed many years later when he formulated his General Theory of Relativity.

Euclid

The "father of geometry" was a Greek mathematician who lived in Alexandria around 300 BCE. In his text, the "Elements" he detailed an axiomatic approach to the known geometry of his time (now called Euclidean geometry): Using rigorous proof to derive theorems from a small number of assumptions. His text became a classic and was used for hundreds of years. The text also contains some discussion of number theory, including the famous proof of the infinitude of primes.

6.2   Relations and Properties

A **cyclic polygon** is one whose vertices lie on the circumference of a circle. It can be proven that every triangle is in fact cyclic: That is, given any triangle, a circle can be drawn through its vertices (see problem set).

Here are some common theorems:

1. **Inscribed Angle Theorem**

   Let A and B be two points on the circumference of a circle. The angle subtended by A and B at the centre of the circle is twice that subtended at a point C on the circumference. See Fig.(6.1).

2. **Tangent-Chord Theorem (Alternate Segment Theorem)**

   Let the triangle ABC be inscribed in a circle and a tangent drawn at point A. Let D be another point on the tangent line such that D and C are on opposite sides of the line AB. Then $\angle DAB = \angle BCA$. See Fig.(6.2).

3. **Intersecting Chords Theorem**

   Let $A, B, C,$ and D be points on the circumference of a circle and X the point of intersection of the lines AC and BD. Then the triangle ABX is similar to triangle DCX. See Fig.(6.3).

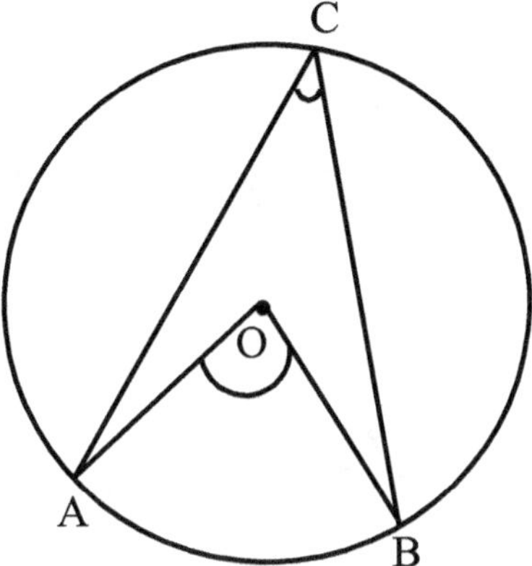

Figure 6.1: Figure illustrating the Inscribed Angle Theorem. Note that the relevant angle at the centre is on the side marked by the letter O.

4. **Tangent-Secant Theorem**

   Let A, B and C be points on the circumference of a circle and let the tangent at A meet the line CB produced at D. Then $(DA)^2 = DB \times DC$. See Fig.(6.4).

6.3 Worked Examples

1. *Prove the Inscribed Angle Theorem*

   Solution:

   Let O be the centre of the circle and the points A, B and C as shown in Fig.(6.1). Let D be a point on CO produced. Now, AOC is an isosceles triangle, and so $\angle AOD = 2\angle ACO$ (why?). Similarly, $\angle BOD = 2\angle BCO$. Hence $\angle AOB = 2\angle ACB$. This completes the proof for the case shown. We leave it to you to consider other locations of C relative to the points A and B. (An important point to note is that in the theorem the "angle subtended at the centre" is the one shown.)

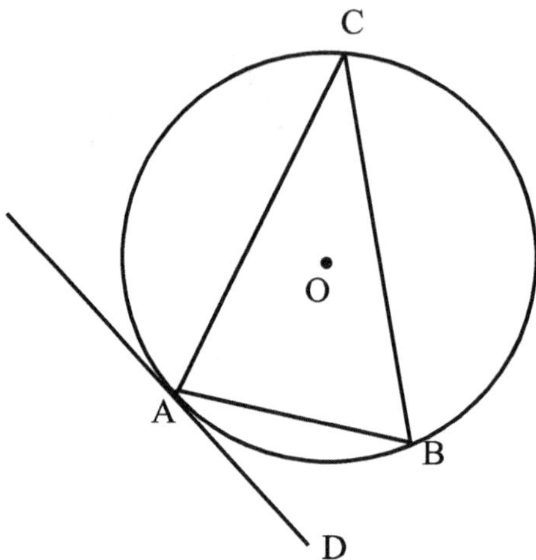

Figure 6.2: Figure illustrating the Tangent-Chord Theorem.

2. *Let AB be a chord of a circle and C be a point on the circumference in one segment and D a point on the circumference in the opposite segment. Prove that (a) The angle ACB does not depend on the exact position of C as long as it stays within one segment, and (b) $\angle ACB + \angle ADB = \pi$.*

Solution:
Let O be the centre of the circle. Then we know from the Inscribed Angle Theorem that $\angle AOB = 2\angle ACB$ regardless of the position of C as long as C is within the segment shown in Fig.(6.1). Using the Inscribed Angle Theorem again, we have $\angle AOB2 = 2\angle ADB$ where $\angle AOB2$ is the angle at the centre which is opposite $\angle AOB$, that is $\angle AOB2 + \angle AOB = 2\pi$. Hence $\angle ACB + \angle ADB = \pi$.

---

3. *Prove that the sum of the exterior angles of a cyclic polygon is 2π regardless of the number of sides .*

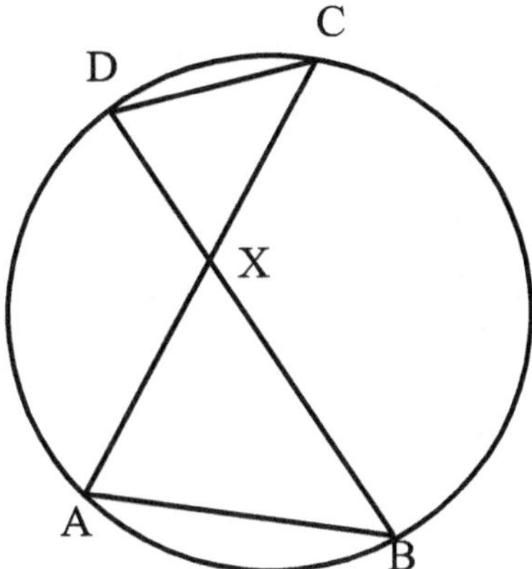

Figure 6.3: Figure illustrating the Intersecting Chords Theorem: Triangle ABX is similar to triangle DCX.

Solution:
Let the polygon have N vertices and draw lines from the centre of the circle to each vertex, dividing the polygon into N triangles as shown in Fig.(6.5). The sum of the interior vertex angles of the polygon is therefore $N\pi - 2\pi$ since each triangle contributes π less the amount at the centre. The sum of the exterior vertex angles is therefore $N\pi - (N\pi - 2\pi) = 2\pi$.

---

4. *Prove the Tangent-Secant Theorem.*

Consider Fig.(6.4). The unique point is A where AD is tangent to the circle of radius R. Let us use that information. Draw the lines OA, OD and OC. From Pythagoras' theorem, $AD^2 = OD^2 - R^2$. On the other hand, we can use the cosine rule on triangle ODC to get $OD^2 - R^2 = CD^2 - 2R(CD)\cos\theta$ where $\theta = \angle OCD$. Next by drawing the line OB we see that $CB = 2R\cos\theta$. Hence putting everything together we get

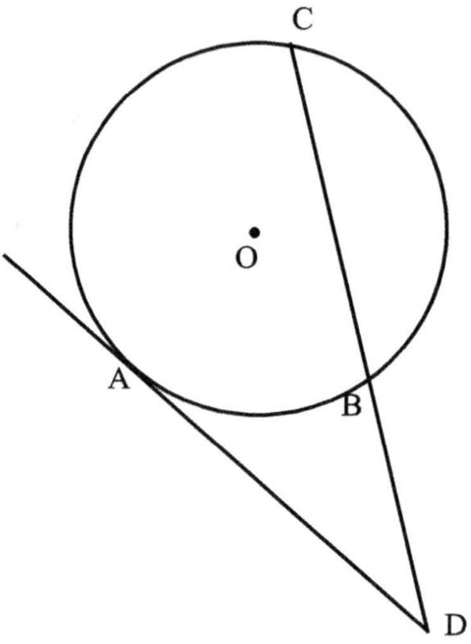

Figure 6.4: Figure illustrating the Tangent-Secant Theorem, $(DA)^2 = DB \times DC$.

$$AD^2 = CD \times (CD - CB) = CD \times BD.$$

An alternate proof of this theorem is in problem $P(10)$.

6.4   Exercises

1. Prove the Theorems listed in Sect.(2).

2. (a) Prove that if a triangle ABC is inscribed in a circle of diameter AB, then $\angle ACB = 90°$.
   (b) Prove that if a right-angled triangle is inscribed in a circle, then the hypotenuse must be the diameter.

3. Prove that the opposite interior angles of a cyclic quadrilateral sum to π.

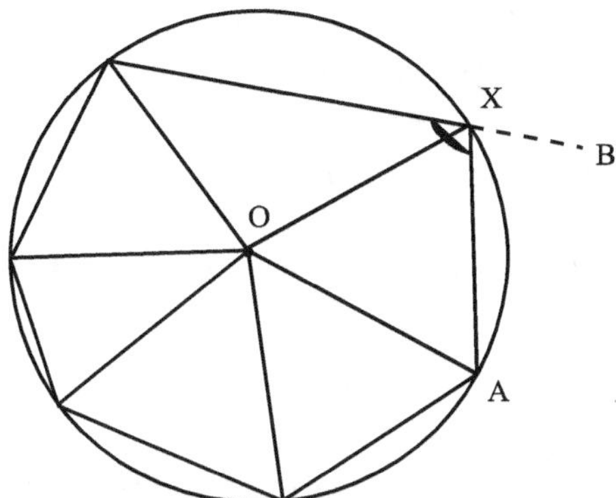

Figure 6.5: Figure for worked example (3). One of the interior vertex angles is marked at X. The straight line produced at X is used to define the exterior vertex angle AXB.

4. Points A, B and C lie on the circumference of a circle. What is the sum $\angle ABC + \angle CAB + \angle BCA$?

5. A right angled triangle is inscribed in a circle. The longest and shortest sides of the triangle are 10 and 5. Determine
   (a) The diameter of the circle.
   (b) The length of the other side of the triangle.

6. The points A, B, C and D lie cyclically on the circumference of a circle with AC and BD being diameters and O the centre. If $\angle CAD = 25°$, find

   (a) $\angle ABD$.   (c) $\angle COD$.
   (b) $\angle BCA$.   (d) $\angle BAD$.

7. $ABCD$ is a cyclic quadrilateral and E is the point outside the circle where BA produced meets CD produced. If $CD = 5$, $DE = 8$ and $EB = 11$ determine the length AE.

8. A circle is inscribed inside a triangle with sides $3, 5$ and 6.
   (a) What is the area of the triangle?
   (b) Find the radius of the circle.

9. The points A, B, C and D lie cyclically on the circumference of a circle. The lines AC and BD intersect at E with $AE = 7$, $EC = 2$, $BE = 6$ and $\angle DEC = 100°$. Find
   (a) The lengths of ED and DC.
   (b) $\angle ABD$.
   (c) The angle between the tangent at C and the line BC.

6.5   Problems

1. A polygon of N sides is inscribed in a circle. What is the sum of all its interior vertex angles?

2. Prove:
   (a) Every triangle may be circumscribed by a circle.
   (b) Every rectangle may be circumscribed by a circle.
   (c) Not every quadrilateral may be circumscribed by a circle.

3. $ABCD$ is a cyclic quadrilateral and E is the point outside the circle where BA produced meets CD produced. If $CD = 5$, $DE = 8$, $EB = 11$ and $\angle CEB = 20°$, determine the lengths BC and AD.

4. AB is the diameter of a circle while BC is tangent to the circle at B. The line AC cuts the circle at D.
   (a) Prove that triangle ADB is similar to triangle BDC.
   (b) Show that $AC \times AD = AB^2$.

5. DC is the diameter of a circle of centre O. The tangent at a point A on the circumference meets DC produced at B. If $CB = 7$ and the radius of the circle is 3, calculate $\angle ADO$.

6. A circle is inscribed inside a triangle with sides $3, 5$ and 6. The sides of the triangle meet the circle at the points X, Y and Z. Find the area of the triangle XYZ.

7. The points A, B, C and D lie cyclically on the circumference of a circle. The lines AC and BD intersect at E with $AE = 7$, $EC = 2$, $BE = 6$ and $\angle ABE = 35°$. Find
   (a) The lengths of ED and DC.

(b) $\angle ADE$.

(c) The angle between the tangent at C and the line EC.

8. A triangle of sides $5, 6$ and 7 is inscribed in a circle. Find the shortest distance from the centre of the circle to the longest side. (Hint: See C3 of the Trigonometry chapter).

9. The points A, B, C and D lie cyclically on the circumference of a circle. The lines AB and AC are of the same length while $BC = 4$. If $\angle BDC = 20°$, find the area of triangle ABC.

10. Consider the cyclic quadrilateral $ABCD$ whose extended lines AD and BC meet at a point P outside the circle.
(a) Show that triangle CPD is similar to triangle APB.
(b) Hence show that $PB \times PC = PA \times PD$.
(c) Now consider sliding the points B and C (keeping the other points fixed) towards each other so that they meet at one point where PB is tangent to the circle. Use this to deduce the Tangent-Secant theorem.

6.6 Challenges

1. Given two non-overlapping circles C_1 and C_2, draw two tangents from the centre of C_1 to the sides of C_2; denote as A and B the points of intersection of the tangents with C_1. Similarly draw two tangents from the centre of C_2 to the sides of C_1; denote the points of intersection of these tangents with C_2 as X and Y. Prove the Eyeball Theorem: The chords XY and AB have the same length.

6.7 Inquire and Investigate

1. Throughout history, there have been several attempts to prove Euclid's fifth postulate. Why did people feel that the fifth postulate was not fundamental?

2. (a) There are geometries which do not adopt Euclid's fifth postulate. Read up on some of these and their possible applications.

(b) The mathematician Riemann discussed curved spaces in general. How did his formalism help Einstein decades later to develop the General Theory of Relativity?

3. Can we empirically investigate if we live in a world with Euclidean geometry?

Chapter 7

Differential Calculus

> *Miracles are not to be multiplied beyond necessity.*
> *Gottfried Leibniz*

7.1 Introduction

Eona heats a square metal sheet uniformly, keeping its square shape as
it expands. If each edge expands at the rate of k mm each second, how
fast is the area of the square surface expanding? We can work this out
as follows: At the instant the edge length is x, the area of one square
side is given by $A = x^2$. When in the time interval Δt, the edge has
expanded to $x + \Delta x$, denote the area by $A + \Delta A$. Hence,

$$
\begin{aligned}
A + \Delta A &= (x + \Delta x)^2 \\
\Delta A &= (x + \Delta x)^2 - x^2 \\
&= 2x\Delta x + (\Delta x)^2 ,
\end{aligned}
$$

so that

$$
\frac{\Delta A}{\Delta t} = 2x\frac{\Delta x}{\Delta t} + \Delta x\frac{\Delta x}{\Delta t} . \tag{7.1}
$$

We now calculate the instantaneous rate of expansion by taking the
limit $\Delta t \to 0$, giving

$$
\frac{dA}{dt} = 2x\frac{dx}{dt} , \tag{7.2}
$$

where $\dfrac{dA}{dt}$ is the usual calculus shorthand for $\displaystyle\lim_{\Delta t \to 0} \dfrac{\Delta A}{\Delta t}$ which denotes the "rate of change of A with respect to t".

Note that as $\Delta t \to 0$, we have $\Delta A \to 0$ but the ratio $\dfrac{\Delta A}{\Delta t}$ tends to a finite limit if $A(t)$ is a smooth function.

To summarise, given a function $A(t)$, one may **differentiate** it to obtain $\dfrac{dA}{dt}$, also referred to as the **derivative of A with respect to t**. If one plots $A(t)$ against t, then $\dfrac{dA}{dt}$ is the slope of the tangent to the curve.

Differential calculus is the branch of mathematics which deals with problems, such as those above, involving some "rate of change". Such problems are common in science as most of the basic equations describing Nature are **differential equations**, that is, equations which contain some derivatives.

For example, Newton's second law of motion, $F = ma$, gives the acceleration, a, of a particle if the forces acting on it are known. If $v(t)$ is the time-varying velocity of the particle, then $a \equiv dv/dt$.

Leibniz

A contemporary of Newton, Leibniz was a German philosopher, lawyer and diplomat who ventured into mathematics late in life. He invented calculus independently of Newton, and although Newton made greater use of it for physical problems, such as determining the planetary orbits, the notation in use nowadays is due to Leibniz. Leibniz also contributed immensely to the development of mechanical calculators, the binary system, aspects of physics and philosophy. His "principle of sufficient reason" and "identity of indiscernibles" are often cited in discourse.

7.2   Relations and Notation

We list here, without proof, some common formulae of differential calculus. They are not all independent; see problem (P2) below.

Let f and g be functions of x, while A, B and n denote constants in the following.

$$\frac{d}{dx} Ax^n = Anx^{n-1}. \tag{7.3}$$

$$\frac{d}{dx} e^x = e^x. \tag{7.4}$$

$$\frac{d}{dx} \ln x = \frac{1}{x}. \tag{7.5}$$

$$\frac{d}{dx} \sin x = \cos x. \tag{7.6}$$

$$\frac{d}{dx} \cos x = -\sin x. \tag{7.7}$$

$$\frac{d}{dx} \tan x = \sec^2 x. \tag{7.8}$$

$$\frac{d}{dx}(f+g) = \frac{df}{dx} + \frac{dg}{dx}. \tag{7.9}$$

$$\frac{d}{dx}(fg) = f\frac{dg}{dx} + g\frac{df}{dx}. \quad \textbf{Product Rule} \tag{7.10}$$

$$\frac{d}{dx}\frac{f}{g} = \frac{gf' - fg'}{g^2}. \quad \textbf{Quotient Rule} \tag{7.11}$$

$$\frac{d}{dx}f(g(x)) = \frac{df}{dg} \times \frac{dg}{dx}. \quad \textbf{Chain Rule} \tag{7.12}$$

$$\frac{df}{dx} = \left(\frac{dx}{df}\right)^{-1}. \tag{7.13}$$

The remarkable relation (7.4) shows the special properties of the constant $e = 2.718....$ In fact Ae^x is the only function which is its own derivative.

Notation: Second derivatives, which express the rate of change of the rate of change, are written $\frac{d}{dx}\left(\frac{dy}{dx}\right) \equiv \frac{d^2y}{dx^2}$.

A convenient short-hand for $\frac{dy}{dx}$ is y', with y'' expressing the second derivative. However if the independent variable is time, t, then the "overdot" notation is usual, that is $\frac{dx}{dt}$ would be written $\dot{x}$ and the second derivative would be $\ddot{x}$.

Caution: Verbally, $\frac{dy}{dx}$ is usually spoken as "dy by dx" and for convenience we will often type it as dy/dx in this book; but remember that $\dfrac{dy}{dx}$ is NOT "dy" divided by "dx". It is simply a notation to express the derivative of y with respect to x, nothing more nor less.

7.2.1 Applications

One application of differential calculus is in finding stationary points of functions. Given a function $f(x)$, the **stationary points** are those for which $\dfrac{df}{dx} = 0$, that is, places where the slope of the function is zero.

The stationary points can be distinguished into **local minima, local maxima or points of inflexion**. Algebraically, their nature is determined by evaluating the second derivative $\dfrac{d^2 f}{dx^2}$ at the stationary point: At local minima the slope of the function increases from left to right and we have $\dfrac{d^2 f}{dx^2} > 0$, while for local maxima the slope of the function decreases, so $\dfrac{d^2 f}{dx^2} < 0$.

Examples are $f(x) = x^2$ which has a local minimum at $x = 0$ and $g(x) = -x^2$ which has a local maximum at $x = 0$. In these two examples the **local extrema** happen to be global extrema as well but this need not be the case in general as you will we see in the worked examples.

Points of inflexion have $\dfrac{d^2 f}{dx^2} = 0$, where the slope does not change sign as we pass the point. An example of this is $y = x^3$ which has a point of inflexion at $x = 0$.

From the definition of $\dfrac{dy}{dx}$ as a limit when $\Delta x \to 0$, we see that if Δx is small but not strictly zero, we may approximate $\dfrac{\Delta y}{\Delta x} \approx \dfrac{dy}{dx}$, so

$$\Delta y \approx \left(\frac{dy}{dx}\right) \times \Delta x, \tag{7.14}$$

a useful method for **approximating the change in a function**. Incidentally, this formula shows that at stationary points, $\Delta y \approx 0$, befitting their name.

7.3 Worked Examples

1. *Differentiate* $\ln\left(\dfrac{e^x}{x} + x\sin x\right).$

Solution:
Let $u = \left(\dfrac{e^x}{x} + x\sin x\right)$ and $y = \ln u$. We wish to evaluate dy/dx. Using the chain rule, $dy/dx = dy/du \times du/dx$. Since $dy/du = 1/u$, we need du/dx. Now, u is the sum of two terms and we can differentiate each separately and add the result. Using the product/quotient rules we have

$$\frac{d}{dx}\frac{e^x}{x} = \frac{xe^x - e^x}{x^2}$$

and

$$\frac{d}{dx}(x\sin x) = \sin x + x\cos x$$

Thus putting everything together, we have

$$\frac{dy}{dx} = \frac{e^x(x-1) + x^2(\sin x + x\cos x)}{xe^x + x^2\sin x}$$

Answer: $\dfrac{e^x(x-1) + x^2(\sin x + x\cos x)}{xe^x + x^2\sin x}.$

---

2. *Determine the stationary points of the curve* $y = x^2 e^{-x}$ *and their nature (maximum, minimum or inflexion). Sketch the curve and clarify the nature of the stationary points.*

Solution:
Differentiating, $dy/dx = 2xe^{-x} - x^2 e^{-x}$. Setting the result to zero gives the stationary points: $2xe^{-x} - x^2 e^{-x} = 0 \Rightarrow x(2 - x) = 0 \Rightarrow x = 0$ or $x = 2$.
The second derivative is $y'' = e^{-x}(2 - 2x) - e^{-x}(2x - x^2) =$

$e^{-x}(2 - 4x + x^2)$; evaluating this at the stationary points gives $y''(0) > 0$ and $y''(2) < 0$. Hence $x = 0$ is a local minimum and $x = 2$ a local maximum.

For sketching the curve it is useful to note that as $x \to \infty$, $y \to 0$ and as $x \to -\infty$, $y \to \infty$. Furthermore $y = 0$ at $x = 0$ but otherwise $y > 0$. See Fig.(7.1). Note from the sketch that $x = 0$ is in fact a global minimum but the curve does exceed the value at $x = 2$ so the latter is only a local maximum.

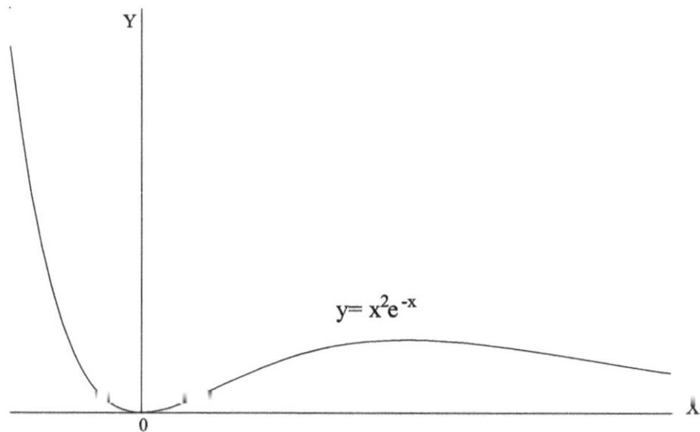

Figure 7.1: The curve $y = x^2 e^{-x}$

Answer: There is a global minimum at $x = 0$ and a local maximum at $x = 2$.

---

3. (a) *Use calculus to determine the approximate change in the value of $y = x + e^{-x} + \ln(x+1)$ when x changes from $x = 0$ to $x = 0.01$.*
   (b) Use a calculator to evaluate $y(0)$ and $y(0.01)$ and compare the difference in values to your result in part (a).
   (c) In a similar manner to part (a), estimate $y(1.02) - y(1)$ and compare using a calculator.
   (d) If x is actually a function of t, and $dx/dt = 3$, calculate dy/dt

when $t = 0$.

Solution:

(a) We need $dy/dx = 1 - e^{-x} + 1/(x+1)$ evaluated at $x = 0$: $y'(0) = 1$. Hence using $\Delta y \approx (dy/dx)\Delta x$ we obtain $\Delta y \approx 0.01$.

(b) $y(0) = 1$ and $y(0.01) \approx 1.01$. The difference is 0.01 which agrees with the result in part (a).

(c) We leave you to verify that the calculus approximation gives $\Delta y \approx 1.132.. \times 0.02 \approx 0.02$ while the exact difference is $(2.083... - 2.061..) \approx 0.02$ as expected.

(d) We have $dy/dt = dy/dx \times dx/dt$. Using the results of part (a), we get $dy/dt = 3$.

Answers: (a) 0.01, (b) 0.01, (c) 0.02, (d) 3.

---

4. *For the curve $y = -x^3 + 2x^2 - 3x + 2$, find the angle that the normal to the curve makes with the x-axis at the point where the curve intersects the y-axis.*

At the y-axis, $x = 0$. We have $y(0) = 2$ and since $y'(x) = -3x^2 + 4x - 3$, so $y'(0) = -3$. That is, the slope of the tangent to the curve at the intercept point is -3. Hence the slope of the normal at that point would be $1/3$. If θ is the angle between the normal and the x-axis, then we know that $\tan\theta = 1/3$, which gives $\theta = 0.32$ radians or $18.4°$.

Answer: $18.4°$.

---

5. *A thin-skinned rubber sphere is inflated by pumping it with water (which is essentially incompressible). If the water is pumped in at the rate of $0.01\ m^3 s^{-1}$, calculate the rate of change of the*

surface area when the radius is 0.5m.

Solution:
The area of the sphere is given by $A = 4\pi R^2$. Differentiating this gives $\dfrac{dA}{dt} = 8\pi R \dfrac{dR}{dt}$. Thus to find the rate of change of area we need the rate of change of its radius which we can obtain from the rate of change of the volume as follows: Since $V = \dfrac{4}{3}\pi R^3$, so $\dfrac{dV}{dt} = 4\pi R^2 \dfrac{dR}{dt} \Rightarrow \dfrac{dR}{dt} = \dfrac{1}{4\pi R^2}\dfrac{dV}{dt}$. Putting everything together, we have $\dfrac{dA}{dt} = \dfrac{2}{R}\dfrac{dV}{dt} = \dfrac{2}{0.5} \times 0.01 = 0.04 \ m^2 s^{-1}$.

Answer: $0.04 \ m^2 s^{-1}$.

---

6. *A student wishes to construct a rectangular box using cardboard and then water-proof its exterior using sticky plastic sheet. Although she has an unlimited amount of cardboard, the total area of plastic sheet available is S m^2. If the water-proofed box must have one at least one square cross-section, determine, in terms of S,*
(a) The linear dimensions of the box that would maximise its volume.
(b) The maximum volume of the constructed box.

Solution:
(a) Denote by x the linear dimension of the square cross-section and h the height of the box. Its exterior surface area is then $A = 2x^2 + 4xh$ while its volume is $V = x^2 h$. Clearly V increases as either x or h increases independently, but then so does the exterior surface area which we know must be bounded, $A \leq S$. So let us first eliminate h from V and write it in terms of A and x: $V = \frac{Ax}{4} - \frac{x^3}{2}$.
We see now that for any given x, V is largest for the largest value of A. So we set $A = S$, the maximum amount of plastic sheet

available for water-proofing, and look for the maximum of V as a function of x.

We have $V'(x) = \frac{S}{4} - \frac{3x^2}{2} = 0 \Rightarrow x = \sqrt{S/6}$ and since $V''(x) = -3x < 0$, the extremum point is a local maximum; a sketch of the curve shows that for $x > 0$ it is actually a global maximum. Thus the volume of the box is maximised for $x = \sqrt{S/6}$ and $h = (S - 2x^2)/4x = \sqrt{S/6}$; that is, the box must be a cube of linear dimension $\sqrt{S/6}$.

(b) $V_{max} = \left(\frac{S}{6}\right)^{3/2}$

Answers: (a) a cube of side $\sqrt{S/6}$, (b) $V_{max} = \left(\frac{S}{6}\right)^{3/2}$.

---

7. *A particle performs simple harmonic motion along the x-axis: Its displacement for $t \geq 0$ is given by $x(t) = 2\sin 3t$.*

   (a) At what time $t > 0$ does the particle return to the origin (i.e. $x = 0$)?

   (b) What is the velocity of the particle when it first returns to the origin for $t > 0$?

   (c) When does the velocity of the particle first become zero for $t > 0$?

   (d) What is the acceleration of the particle when it reaches the situation in part (c)?

   (e) When is the acceleration of the particle first a maximum for $t > 0$? What is the value of the maximum acceleration?

Solution:

(a) $x = 0 \Rightarrow \sin 3t = 0 \Rightarrow 3t = n\pi$ with $n = 0, 1, 2....$ So the first return is at $t = \pi/3$.

(b) Velocity is the rate of change of displacement, $v(t) = dx/dt = 6\cos 3t$. So $v(\pi/3) = 6\cos \pi = -6$. (Note the minus sign indicates the particle is now traveling in the negative x direction).

(c) $v = 0 \Rightarrow \cos 3t = 0 \Rightarrow 3t = (n+1)\pi/2$ with $n = 0, 1, 2....$ So $v = 0$ first occurs when $t = \pi/6$.

(d) Acceleration is the rate of change of velocity, $a(t) = dv/dt = -18\sin 3t$ and $a(\pi/6) = -18$.

(e) Since $a(t) = -18\sin 3t$ and since the sine function varies between -1 and 1, The maximum value of a is 18 and this occurs when $\sin 3t = -1$, that is when $3t = 3\pi/2 \Rightarrow t = \pi/2$.

Answers: (a) $t = \pi/3$, (b) $v = -6$, (c) $t = \pi/6$, (d) $a = -18$, (e) $t = \pi/2$, $a = 18$.

7.4   Exercises

1. For question E(3) of Chapter (4) determine the velocity of the projectile in the x and y directions as a function of time.

2. Verify the identities (7.10-7.13) for the cases $f(x) = x^n$ and $g(x) = x^m$.

3. Differentiate each of the following expressions with respect to x:

(a) $3x^2 + 2x^{-3}$.

(b) $\ln(2x+1)$.

(c) $2\sin 3x$.

(d) $3\cos^2 x$.

(e) $3xe^{2x}$.

(f) $\dfrac{2+3x}{x+5}$.

4. Suppose that the variable x is a function of time, and y is another variable defined by $y = f(x)$ where $f(x)$ is given by the expressions of the previous exercise. Calculate, for each case of the previous exercise, $\dfrac{dy}{dt}$ at $x = 0$ if $\dfrac{dx}{dt} = 2$.

5. Differentiate each of the following expressions with respect to x:

(a) $3x^2(1+x) + \dfrac{2}{x}$.

(b) $\sqrt{2+3x}$.

(c) $\dfrac{\sqrt{2+3x}}{x^3+5}$.

(d) $\sin 2x + x^2 \cos x + x \tan 5x$.

(e) $x^2 e^x + \ln(1+x)$.

(f) $\sin(\ln x)$.

6. The von Bertalanffy model for tumour growth is given by the equation.

$$\frac{dV}{dt} = aV^p - bV^q,$$

where V is the size of the tumor, t the time and a, b, p, q are positive constants.

(a) The two terms on the right hand side of the equation represent growth and degradation. Identify those respective terms.

(b) For which range of V values will the tumour grow in size?

(c) Optional: Read about a practical application of such tumour growth models.

7. For the parabola $y = ax^2 + bx + c$,

(a) Evaluate the first and second derivatives with respect to x.

(b) Hence deduce that there is always one turning point and explain its nature.

(c) Show that the results of part (b) may also be obtained without using calculus, by completing the square.

8. Determine the stationary points, if any, of the following curves and their nature (maximum, minimum or inflexion). Where possible, sketch the curves to decide if the extrema are local or global.

(a) $y = 2x + \dfrac{1}{1+x}$.

(b) $y = \sqrt{2+3x^2}$.

(c) $y = (x+1)^2 e^{-x}$.

(d) $y = \sin x^2 + x^2 \quad (x^2 \le \pi)$.

(e) $y = \ln(x^2 + x + 3)$.

9. Find the equation of the tangent and normal to each of the following curves at the point $x = 1$:

(a) $y = x^2 - 3x + 5$.

(b) $y = x^3 + x^2 + 2x + 1$.

10. (a) Find the range of values of x for which y decreases as x increases for the curves in the previous exercise.
    (b) Find the range of values of x for which the slope is increasing (for increasing x) for the curves in exercise (9).

11. A biology student finds that the time change of population, $\dfrac{dx}{dt}$, of bacterial cells in his sample is well described by the logistic equation $\dfrac{dx}{dt} = Ax(B - x)$, with A, B positive constants.
    (a) Determine the stationary points of the population growth.
    (b) What happens when $x > B$?
    (c) Sketch a curve of $x(t)$.
    (d) Suggest an interpretation of the constant B.
    (e) Optional: Discuss another real world application of the logistic equation.

12. A thin-skinned rubber sphere is inflated by pumping it with water (which is essentially incompressible). If the radius of the sphere increases at the rate of k cm/s, calculate, in terms of k, the rate of change of the following (Express your answer in terms of the instantaneous radius):
    (a) The circumference of a great circle on the sphere.
    (b) The surface area of the sphere.
    (c) The volume of the sphere.

13. If the radius of an inflating sphere changes from 10 cm to 10.1 cm, use calculus to estimate the corresponding change in the
    (a) Circumference of a great circle.
    (b) The surface area of the sphere.
    (c) The volume of the sphere.
    (d) Check the accuracy of your answers above by calculating the relevant quantities exactly.

14. Verify the identity (7.13) for the cases $y = x^2$ and $y = \ln x$.

7.5 Problems

1. For the cubic curve $y = ax^3 + bx^2 + cx + d$, $(a \neq 0, b \neq 0)$
   (a) Evaluate the first and second derivatives with respect to x.
   (b) Hence deduce the condition for the cubic curve to have no turning points.
   (c) Under what conditions will the cubic curve have two turning points?
   (d) Show that when there is exactly one stationary point, it must be a point of inflexion.

2. Deduce the following relations from the other relations listed in Sect.(2):

   (a) $\dfrac{d}{dx} \cos x = -\sin x$ (Hint: Use $\sin^2 x + \cos^2 x = 1$).

   (b) $\dfrac{d}{dx} \tan x = \sec^2 x$.

   (c) $\dfrac{d}{dx} \sec x = \sec x \tan x$.

   (d) $\dfrac{d}{dx} \csc x = -\csc x \cot x$.

   (e) $\dfrac{d}{dx} \cot x = -\csc^2 x$.

   (f) $\dfrac{d \ln x}{dx} = \dfrac{1}{x}$.

   (g) $\dfrac{d}{dx} \dfrac{f}{g} = \dfrac{gf' - fg'}{g^2}$.

3. Differentiate each of the following expressions with respect to x:

   (a) $(\tan x)(\sin^2 x)$.

   (b) xe^{x^2}.

   (c) $\dfrac{\sin x}{1 + 3x + x^2}$.

   (d) $\sin(\ln(x^2 + 1))$.

   (e) $\ln(\cos x^2)$.

   (f) $\dfrac{(\ln 3x)^2}{x^2}$.

4. Determine the approximate change in the value of each expression in the previous problem when x changes from $x = 1$ to $x = 1.01$.

5. If the volume of an inflating sphere changes from 50 cm^3 to 51 cm^3,
   (a) Use calculus to estimate the change in its radius.
   (b) Check the result in part (a) by computing the corresponding radii for the two spheres exactly.

6. Find the derivatives of $\log_{10} x$ and a^x for $a > 0$.

7. Determine the value of the second derivative of each of the following at $x = 1$:

   (a) $3x^2(1+x) + 2\ln x$.

   (b) $\sin \ln x$.

   (c) $\sqrt{2 + 3x^2}$.

   (d) $x \tan x$.

   (e) $x^2 e^x + \ln(1+x)$.

   (f) $\dfrac{2 + 3x}{x + 1}$.

8. (a) For the curve $y = -x^3 + 2x^2 - x + 2$, find the equations of all tangents to the curve which have slope -5.
   (b) Determine the equations of the normals to the curve at the points determined in part (a).

9. A particle performs damped harmonic motion along the x-axis: Its displacement for $t \geq 0$ is given by $x(t) = 2e^{-t}\sin(12t)$.
   (a) At what time $t > 0$ does the particle first return to the origin (i.e. $x = 0$)?
   (b) What is the velocity of the particle when it first returns to the origin for $t > 0$?
   (c) When does the velocity of the particle first become zero for $t > 0$?
   (d) What is the acceleration of the particle when it reaches the situation in part (c)?
   (e) When is the acceleration first zero for $t \geq 0$?
   (f) Sketch the $x(t)$ curve.

10. The top of a swimming pool has a rectangular cross-section with length L and width W. The shallow end of the pool has depth h and the floor of the pool slopes at a constant angle towards the deep end of the pool which has depth H. If water is drained from the bottom of the pool at the rate of v m^3s^{-1}, determine the rate of change of water height in the pool when the pool is almost full.

11. A farmer wants to fence off a rectangular section of his farm but the material he has can only cover a boundary of length L. Determine the dimensions of the rectangular section (in terms of L) which would
    (a) Maximise the enclosed area.
    (b) Minimise the enclosed area.

12. The boundaries of a school field consist of the curved part of a semi-circle and its diameter. The curved part has radius R. A rectangle is to be drawn with all its vertices on the boundary of the field. Determine the rectangles of largest and smallest areas that can be so drawn.

13. A right-angled triangle has area A and one acute angle θ. A square is inscribed in the triangle with exactly one vertex on the hypotenuse.
    (a) Determine the area of the square in terms of A and θ.
    (b) If θ is allowed to vary while A is fixed, use calculus to determine the maximum possible area of the square.
    (c) Can you obtain the answer to part (b) without using calculus? (Hint: Try a symmetry argument).

7.6 Challenges

1. Determine the slope of the curve $xy^2 + y - x^2 - 3x - 1 = 0$ at the point(s) with $x = 1$.

2. A thin-skinned rubber sphere is inflated by pumping it with water (which is essentially incompressible). The surface of the sphere has a number of non-intersecting closed curves painted on it. If

the fractional rate of increase of the radius of the sphere is $C\ s^{-1}$, calculate the fractional rate of change of the total area enclosed by the painted curves. (Note: $C = \frac{1}{R}\frac{dR}{dt}$).

3. Given a circle, determine the maximum possible area of a triangle inscribed in it.

4. A right-angled triangle has area A and one acute angle θ. A circle is inscribed in the triangle. If θ is allowed to vary while A is fixed, determine the maximum possible area of the circle.

7.7   Inquire and Investigate

1. (a) Historically, several different notations have been used, and some continue to be used, for the derivative of a function. Discuss the different notations, their advantages and disadvantages.
   (b) What is the origin of the word "calculus"?

2. (a) Consider the function $y = |x - 1|$. What would be its slope at $x = 1$?.
   (b) Is it possible for a "curve" to have no points at which it is differentiable?

3. Let $f(z)$ be a function of a complex variable z. Is it possible to define the derivative of such a function? How would calculus for functions of complex variables differ from calculus for functions of real variables?

Chapter 8

Integral Calculus

If I have been able to see further,
it was only because I stood on the shoulders of giants.
Isaac Newton

8.1 Introduction

We know how to calculate the areas of simple figures such as a square or circle. How then do we calculate the areas of more complicated figures, especially those that have no symmetry? Consider for example the continuous function $y = f(x) > 0$ in the region $a \leq x \leq b$. What is the area bounded by the x-axis and the curve in that region? For example, can you deduce the area for a parabola, $f(x) = x^2$?

You could imagine dividing the region $a \leq x \leq b$ into thin strips of width Δx and approximating each strip by a rectangle of height $f(x_i)$ where x_i is a point within that strip. Then by adding up all the rectangular areas and taking the limit $\Delta x \to 0$ you should be able to obtain the required area, see Fig.(8.1).

Formally, the area so obtained is denoted by the **definite integral**

$$A = \int_a^b f(x)dx \,. \tag{8.1}$$

But is there a way to calculate A directly without doing the summation of thin strips and then taking limits? That is, how does one evaluate

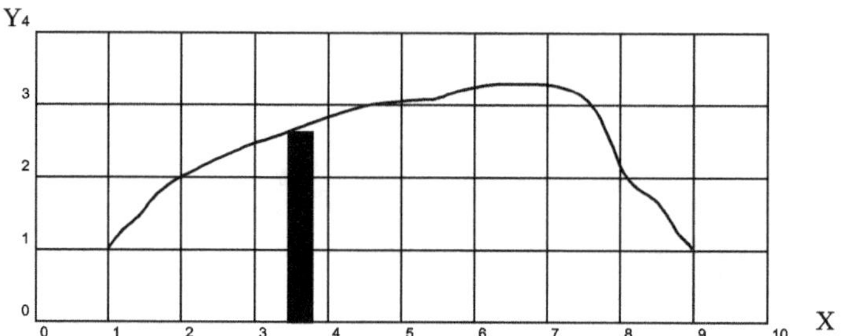

Figure 8.1: The area under the curve can be approximated by dividing it into narrow strips of width Δx, one of which is shown.

the integral in (8.1) directly? The answer is provided by the **Fundamental Theorem of Calculus** which links integral calculus with differential calculus:

$$\int_a^b f(x)dx = F(b) - F(a)\,, \tag{8.2}$$

where $F(x)$ is a function that satisfies

$$\frac{dF(x)}{dx} = f(x)\,. \tag{8.3}$$

In the above example we illustrated the idea of a definite integral by calculating areas, and we imposed the condition $f(x) > 0$ for simplicity. However the theorem (8.2) holds for general $f(x)$.

We see from (8.2-8.3) that the process of "integration" is essentially the reverse of differentiation,

$$\int_a^b \frac{dF(x)}{dx}\,dx = F(b) - F(a)\,. \tag{8.4}$$

If we omit the limits of the integration region, we have an **indefinite integral** given by

$$\int f(x)dx = F(x) + C\,, \tag{8.5}$$

where (8.3) still holds and C is a constant of integration which can be fixed once we have more information about the problem.

Since, as we saw in the last chapter, many equations in science involve rates of change (derivatives), the process of integration is often needed to obtain an expression for the quantity that is changing. For example, while Newton's Second Law gives us the acceleration of an object, to get its velocity and displacement requires us to perform some integration; this is illustrated in some of the practice questions.

Newton

The epitome of genius, Newton formulated the laws of mechanics, the law of universal gravitation, and developed the necessary mathematical tools, such as calculus, to investigate those laws while still a student at Cambridge. However he did not publicise his results until many years later. He also made original contributions to optics and helped solve a problem with the English coinage. Newton achieved legendary status within his lifetime, not only among his peers, but also among the general public.

8.2 Relations and Properties

We list here, without proof, some common formulae of integral calculus. Let f and g be functions of x, while a, b and n denote constants. In the formulae for indefinite integrals below, C is a constant of integration.

$$\int (a + bx)^n \, dx \;=\; \frac{(a + bx)^{n+1}}{b(n + 1)} + C, \quad n \neq -1 \, . \qquad (8.6)$$

$$\int \frac{1}{a + bx} \, dx \;=\; \frac{1}{b} \ln(a + bx) + C \, . \qquad (8.7)$$

$$\int \sin(a + bx) \, dx \;=\; \frac{-1}{b} \cos(a + bx) + C \, . \qquad (8.8)$$

$$\int \cos(a + bx) \, dx \;=\; \frac{1}{b} \sin(a + bx) + C \, . \qquad (8.9)$$

$$\int e^{(a+bx)} \, dx \;=\; \frac{1}{b} e^{(a+bx)} + C \, . \qquad (8.10)$$

$$\int (f + g) \, dx \;=\; \int f \, dx + \int g \, dx \, . \qquad (8.11)$$

It is useful to note the following identity

$$\int_a^c f(x)dx = \int_a^b f(x)dx + \int_b^c f(x)dx. \qquad (8.12)$$

For calculating areas using (8.1), note that if $f(x) < 0$ within a region $x_1 \le x \le x_2$, the integral in that region would give a negative value and the area is then the negative of the integral.

One may also evaluate the area between a curve and the y-axis. In this case the integral would be $\int_{y_1}^{y_2} x\, dy$.

8.3   Worked Examples

1. *Evaluate* $\displaystyle\int_0^\pi \left(3x^2\left(1 + \frac{1}{x}\right)^2 + 3\sin 3x + \frac{1}{1+3x} + 3e^{-2x} \right) dx$.

Solution:

Using (8.11), the integral may be conveniently evaluated as four separately pieces which can then be added together:

$$\begin{aligned}
\int_0^\pi 3x^2\left(1 + \frac{1}{x}\right)^2 dx &= \int_0^\pi (3x^2 + 6x + 3)\, dx \\[2mm]
&= \left. (x^3 + 3x^2 + 3x) \right|_0^\pi \\[2mm]
&= \pi^3 + 3\pi^2 + 3\pi . \qquad (8.13)
\end{aligned}$$

$$\begin{aligned}
\int_0^\pi 3\sin 3x\, dx &= \left. -\cos 3x \right|_0^\pi \\[2mm]
&= -(\cos 3\pi - \cos 0) \\[2mm]
&= 2. \qquad (8.14)
\end{aligned}$$

$$\begin{aligned}
\int_0^\pi \frac{dx}{1+3x} &= \left. \frac{\ln(1+3x)}{3} \right|_0^\pi \\[2mm]
&= \frac{\ln(1+3\pi)}{3} - \frac{\ln(1)}{3} \\[2mm]
&= \frac{\ln(1+3\pi)}{3}. \qquad (8.15)
\end{aligned}$$

$$\int_0^\pi 3e^{-2x}\ dx = \frac{-3}{2}e^{-2x}\Big|_0^\pi$$

$$= \frac{-3}{2}(e^{-2\pi} - 1). \tag{8.16}$$

Answer: $\pi^3 + 3\pi^2 + 3\pi + \dfrac{7}{2} + \dfrac{\ln(1 + 3\pi)}{3} - \dfrac{3}{2}e^{-2\pi} \approx 74.3$

2. *The tangent to a curve $y = f(x)$ at the point (x, y) has slope $x + e^{2x}$. Determine the equation of that curve which includes the point $(0, 3)$.*

Solution:
We have

$$\frac{dy}{dx} = x + e^{2x}$$

$$\Rightarrow y(x) = \int \left(x + e^{2x}\right)\ dx = \frac{x^2}{2} + \frac{e^{2x}}{2} + C$$

where C is the integration constant that can be determined from the condition $y(0) = 3 \Rightarrow C = 5/2$.

Answer: $y(x) = \dfrac{x^2}{2} + \dfrac{e^{2x}}{2} + \dfrac{5}{2}.$

3. *Determine the area bounded by the $x - $ axis and the curve $y = x(x - 1)(x - 3)$ between each pair of consecutive zeros of the curve.*

Solution:
Firstly, note that $y > 0$ for $0 < x < 1$ and $y < 0$ for $1 < x < 3$, so the integral in the first domain will give us the area but the

integral over the second domain will give us the negative of the area. We have

$$\int x(x-1)(x-3)\,dx \;=\; \int (x^3 - 4x^2 + 3x)\,dx$$

$$= \; \frac{x^4}{4} - \frac{4x^3}{3} + \frac{3x^2}{2} + \text{constant}\,,$$

so

$$\int_0^1 x(x-1)(x-3)\,dx \;=\; \left(\frac{x^4}{4} - \frac{4x^3}{3} + \frac{3x^2}{2}\right)\Bigg|_0^1 = \frac{5}{12}$$

and

$$\int_1^3 x(x-1)(x-3)\,dx \;=\; \left(\frac{x^4}{4} - \frac{4x^3}{3} + \frac{3x^2}{2}\right)\Bigg|_1^3 = \frac{-8}{3}$$

Answers: 5/3 between the first two zeros and 8/3 between the last two.

---

4. *Find the area bounded between the curves $y = 2x^2$ and $y = 3\sqrt{x}$.*

Solution:

Let us find the intersection points of the two curves. Setting $2x^2 = 3\sqrt{x}$, gives $x = 0$ or $x = \left(\frac{3}{2}\right)^{2/3}$. Sketching the two curves shows that the bounded area is the difference of two areas: The area below $y = 3\sqrt{x}$ minus the area below $y = 2x^2$ evaluated between the intersection points.

So $A = \displaystyle\int_0^a (3\sqrt{x} - 2x^2)\,dx$, where $a = \left(\frac{3}{2}\right)^{2/3}$. We leave the computation to you.

Answer: 3/2.

5. *Sand flows into a vertically held hollow right circular cone at its large end and flows out at the narrow end. At time $t > 0$, sand is flowing in at the rate of $10t$ $cm^3 s^{-1}$ and flows out at the rate of t^2 $cm^3 s^{-1}$. If at time $t = 0$ the cone holds 10 cm^3 of sand,*

   (a) Determine the volume of sand, $V(t)$, in the cone at time t.

   (b) Sketch the curve $V(t)$ for $t \geq 0$, locating its local maxima and minima.

   (c) At what value of $t > 0$ will the the amount of sand in the cone reach its maximum value V_{max}? What is the value of V_{max}?

   (d) Determine the value of $t \geq 0$ when the amount of sand in the cone is at a local minimum.

   (e) Beyond which time does the amount of sand in the cone go below the local minimum value found in part (b)?

   (f) Will there be a time when there is no sand in the cone? Explain.

   (g) If the volume of sand in the cone is given by $V = \frac{h^3}{20} cm^3$ where h is the height of the sand in the cone, calculate the rate of change of the height of the sand when $t = 5$ s.

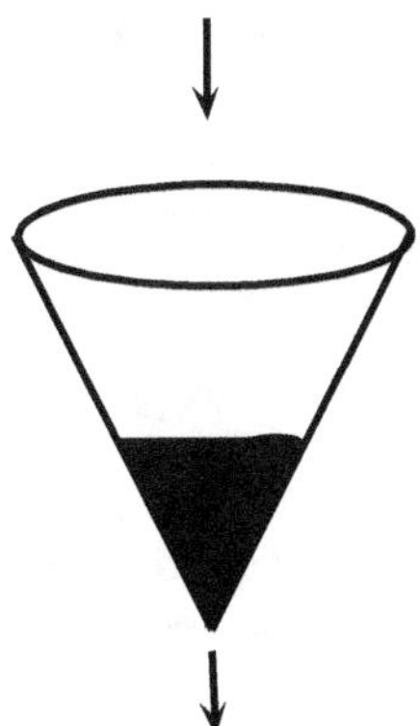

Figure 8.2: The cone filling and emptying.

Solution:

(a) We have $\dfrac{dV}{dt} = 10t - t^2 \Rightarrow V(t) = 5t^2 - \dfrac{t^3}{3} + C$ where the constant C is determined by the given condition $V(0) = 10 \Rightarrow C = 10$. Thus $V(t) = 5t^2 - \frac{t^3}{3} + 10$.

(b) The stationary points satisfy $0 = \dfrac{dV}{dt} = 10t - t^2 \Rightarrow t = 0$ or $t = 10$. From $\dfrac{d^2V}{dt^2} = 10 - 2t$ one deduces that $t = 0$ is a local minimum (why?) and $t = 10$ a local maximum (why?). Note that as $t \to +\infty$, $V(t) \to -\infty$ and as $t \to -\infty$, $V(t) \to +\infty$. With this information the curve can be sketched, see Fig.(8.3).

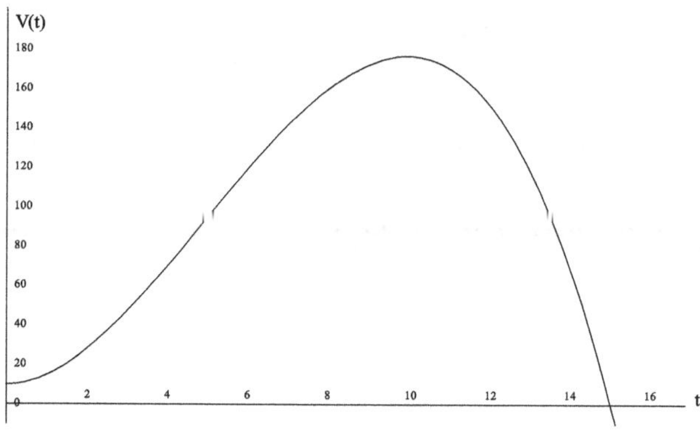

Figure 8.3: Sketch of $V(t)$.

(c) From parts (a) and (b) we find $V_{max} = V(10) = 530/3$ at $t = 10$.

(d) From parts (a) and (b) we find that a local minimum is reached when $t = 0$.

(e) At $t = 0$, $V = 10$. From the curve we see that this value will be reached later also. Indeed solving $V(t) = 10$ gives $t = 0$ or $t = 15$. For $t > 15$, the amount of sand in the cone drops below its local minimum value at time $t = 0$.

(f) Yes. From the curve we see that at some time beyond $t = 15$ one will have $V = 0$.

(g) Differentiating $V = \dfrac{h^3}{20}$ with respect to t gives $\dfrac{dV}{dt} = \dfrac{3h^2}{20}\dfrac{dh}{dt}$ and so

$$\frac{dh}{dt} = \frac{20}{3h^2}\frac{dV}{dt}. \tag{8.17}$$

From previous expressions (see part (a)), we know V and dV/dt at $t = 5$. Furthermore since $h = (20V)^{1/3}$ (why?), so the value of h at $t = 5$ is known. Putting everything together in (8.17) gives $dh/dt = 1.1$.

Answers: (a) $V(t) = 5t^2 - \frac{t^3}{3} + 10$, (c) $V_{max} = V(10) = 530/3$ at $t = 10$, (d) $t = 0$, (e) $t = 15$, (f) Yes, (g) 1.1.

---

6. *If $y = x \ln x$, find dy/dx. Hence evaluate $\int dx \ \ln x$.*

Solution:

We have $\dfrac{d}{dx}(x \ln x) = x \cdot \dfrac{1}{x} + 1 \cdot \ln x = 1 + \ln x$.

The above result may be re-written as

$\ln x = \dfrac{d}{dx}(x \ln x) - 1 = \dfrac{d}{dx}(x \ln x - x)$. Hence from (8.5) we get

$$\int \ln x \ dx = (x \ln x - x) + C.$$

8.4   Exercises

1. Sketch each of the following curves and then find the area between the curves and the x-axis within the range indicated:

(a) $\sin x$, $0 \le x \le \pi/2$.

(c) $\dfrac{1}{x}$, $1 \le x \le 2$.

(b) e^x, $-\infty \le x \le 0$.

(d) $\dfrac{1}{x^2}$, $1 \le x \le \infty$.

(e) $\sqrt{x}$, $0 \le x \le 1$.

(f) x^2, $0 \le x \le 1$.

2. (a) Evaluate $\displaystyle\int_0^{2\pi} \sin x \, dx$.

   (b) Find the area between the x-axis and $y = \sin x$ for $0 \le x \le 2\pi$.

   (c) Why is the answer to part (b) different from that in part (a)?

3. Integrate each of the following with respect to x:

   (a) $3x^2(1 + x) - \dfrac{2}{x}$.

   (b) $\sqrt{2 + 3x}$.

   (c) $2\sin 2x - 3\sec^2 x$.

   (d) $\dfrac{2}{5 + 3x}$.

   (e) $2e^{3x} - (1 + x)^3$.

   (f) $7\sin 2x \cos 2x$.

   (g) $\left(\dfrac{1}{e^{2x}} - \dfrac{1}{\sqrt{2 + 3x}} \right)$.

4. Evaluate each of the following definite integrals to two decimal places:

   (a) $\displaystyle\int_1^2 dx \left(3x^2(1 + x) + \dfrac{2}{x} \right)$.

   (b) $\displaystyle\int_0^1 \sqrt{2 + 3x} \, dx$.

   (c) $\displaystyle\int_{-1}^1 \dfrac{2}{5 + 3x} \, dx$.

   (d) $\displaystyle\int_0^{\pi/4} dx \left(2\sin 2x + 3\sec^2 x \right)$.

   (e) $\displaystyle\int_0^1 dx \left(2e^{3x} + (1 + x)^3 \right)$.

   (f) $\displaystyle\int_0^2 dx \left(\dfrac{1}{e^{2x}} + \dfrac{1}{\sqrt{2 + 3x}} \right)$.

   (g) $\displaystyle\int_0^{\pi} \sin 2x \cos 2x \, dx$.

5. The normal to a curve $y = f(x)$ at the point (x, y) has slope $\dfrac{x}{1 - x}$. Determine the equation of that curve which includes the point $(1, 2)$.

6. A thin-skinned rubber sphere is inflated by pumping it with water (which is essentially incompressible). The radius R of the sphere increases at the rate of 0.5 cm/s. If $R = 10$ cm at $t = 0$, determine the radius of the sphere at $t = 5$ s.

7. Show that the area bounded by the curve $y = \sin^2 x$ and the x-axis, between the points $x = 0$ and $x = 2\pi$, is less than 2π.

8. Determine the area bounded by the x-axis and the curve $y = x(x-1)(x-3)$ between consecutive turning points of the curve.

9. The acceleration of a particle moving along the (vertical) y-axis is constant, $a(t) \equiv \ddot{y} = -g$. If its velocity $v(t) = \dot{y}$ and displacement $y(t)$ at $t = 0$ are $v(0) = u > 0$ and $y(0) = 0$, determine
   (a) $v(t)$ and sketch the corresponding curve.
   (b) $y(t)$ and sketch the corresponding curve.
   (c) The time when the particle reaches the maximum height.
   (d) The maximum height reached by the particle.
   (e) The time when the particle returns to $y = 0$.
   (f) The velocity of the particle when it returns to $y = 0$.
   (g) Does this problem have a physical application?

10. If $\displaystyle\int_1^6 f(x)dx = 20$ and $\int_1^8 f(x)dx = 15$,

   (a) Find $\displaystyle\int_6^8 f(x)dx$.
   (b) What can you say about the sign of $f(x)$ for $1 \le x \le 6$?
   (c) What can you say about the sign of $f(x)$ for $6 \le x \le 8$?

8.5   Problems

1. Find the area between the following curves and the x-axis within the range indicated:

   (a) $\sin^2 x$, $0 \le x \le \pi/2$.

   (b) $\cos^2 x - \sin^2 x$, $0 \le x \le \pi/2$.

   (c) $\sec^2 x$, $0 \le x \le \pi/4$.

   (d) $\dfrac{1}{\sqrt{x}}$, $0 \le x \le 1$.

2. If $\displaystyle\int_0^p (2 + 3x)\, dx = 5$, determine p.

3. Integrate the following with respect to x:

(a) $\displaystyle\frac{2x + 1}{x^2 + 3x + 2}$.

(b) $\displaystyle\frac{5x^2 + 2x + 1}{x^2 + 3x + 2}$.

(c) $\displaystyle\frac{5x^2 + 2x + 1}{x + 1}$.

(d) $\displaystyle\frac{5x^2 + 2x + 1}{x^2 + 2x + 1}$.

(e) $\displaystyle\frac{x^4 + 1}{x^3 + x}$.

4. Evaluate the definite integral of each expression in the previous problem for end points at $x = 1$ and $x = 2$. Give your answer to two decimal places.

5. Sketch the curves in each case below and determine the area bounded by them.
(a) $y = (x - 1)^2 + 1$ and $y = 4$.
(b) $y = (x - 1)^2 - 1$ and $y = 0$.
(c) $y = (x - 1)^2 + 1$, $y = 2$ and $y = 4$.
(d) $y = 2 - 3x^2$ and $y = x^2$.
(e) $y = x^2$, $y = 1/x^2$ and $y = 0$ for $0 \le x \le \infty$.

6. Determine the area bounded by the curves in each case below:
(a) $y = (x - 1)^2 + 1$ and $y = -(x - 1)^2 + 2$. (Hint: There is more than one way to do this).
(b) $xy = 1$, $y = x$, $y = 0$ for $0 \le x \le 10$.

7. Find the area bounded by the curves and the **y**-axis for each case below:
(a) $y = \ln x$, for $0 \le y \le 1$.
(b) $y = \sqrt{x}$, for $1 \le y \le 2$.
(c) $y = 4x^2$ and $y = 10 - x$ for $x \ge 0$.

8. The tangent to a curve $y = f(x)$ at the point (x, y) has slope $\dfrac{x}{1 - x}$. Determine the equation of that curve which includes the point $(0, 3)$.

9. For some curves described by the equation $y = f(x)$, the gradient $f'(x)$ is given below for $0 \leq x \leq 1$. In each case the curve has a stationary point at $y = 1$. Determine the equation of each curve. (If in any case the answer happens not to be unique, determine all solutions.):

   (a) $(1 + x) - \dfrac{4}{x + 1}$.

   (b) $7 \sin 2x \cos 2x$.

10. A thin-skinned rubber sphere is inflated by pumping it with water (which is essentially incompressible). Water is pumped in at the rate of $5 \ cm^3 s^{-1}$. If the volume of the sphere is $100 cm^3$ when $t = 5$, determine

    (a) The radius of the sphere at $t = 1$.

    (b) The rate of change of the radius at $t = 1$.

    (c) The rate of change of the surface area at $t = 1$.

11. A particle moves along the x-axis with velocity $v(t) = 2 - t^2 + 3t$.

    (a) If $x = 2$ when $t = 0$, determine $x(t)$.

    (b) At what time $t_1 > 0$ does the particle re-appear at $x = 2$?

    (c) At what time t_2 is the particles' velocity instantaneously zero?

    (d) Sketch $x(t)$ versus t for $t > 0$.

    (e) What distance has the particle travelled between times t_1 and t_2?

12. A particle moves along the the x-axis with acceleration $a(t) \equiv \ddot{x} = t - e^{-t}$. If its velocity $v(t) \equiv \dot{x}$ and displacement $x(t)$ at $t = 0$ are $v(0) = 2$ and $x(0) = 0$, determine

    (a) $v(t)$ and sketch the corresponding curve.

    (b) $x(t)$ and sketch the corresponding curve.

    (c) The number of times the particle changes direction.

    (d) The number of times the particle appears at $x = 0$.

    (e) The total displacement of the particle between $t = 0$ and $t = 3$.

13. A particle's displacement is given by $x(t) = 2 \sin t$ metres.

    (a) What is the displacement of the particle between times $t = 0$ and $t = 3\pi/2$?

    (b) What is the distance travelled by the particle between times

$t = 0$ and $t = 3\pi/2$?

(c) Why are the answers to parts (a) and (b) different?

14. For the worked example (5), let $t = T$ be the time when there is no more sand in the cone.

(a) Show by direct substitution, and using the sketch of $V(t)$, that $15 < T < 15.5$.

(b) Writing $T = 15 + x$, and setting $V(T) = 0$, find an approximate linear equation for x. (That is, ignore quadratic and higher order terms in x. Why is this justified? See similar problems in the Supplementary chapter.)

(c) Hence deduce an approximate value for T.

15. It is known that $y(x) = x + \ln(x + 1)$ and that x is a function of t. If $dx/dt = 2t$ and $x = 1$ at $t = 0$, calculate dy/dt at $t = 2$.

16. Using the identity $\sin^2\theta + \cos^2\theta = 1$, and the symmetry of those functions, deduce, without explicit computation, that
$\int_0^1 \sin^2(2\pi x)\,dx = \int_0^1 \cos^2(2\pi x)\,dx = 1/2$.

17. Given a curve $y = f(x)$, its length between two points may be calculated using the formula
$L = \int_a^b dx\sqrt{1 + (y')^2}$ where $y' = dy/dx$.

(a) Verify the formula for the straight line $y = mx$ between $0 \le x \le 1$.

(b) Calculate the length of the curve $y = x^{3/2}$ between $0 \le x \le 1$.

(c) Optional: Study the derivation of the formula.

18. Given a curve $y = f(x) > 0$ between the points $x = a$ and $x = b$, rotating it about the x-axis creates a three dimensional figure whose curved surface area S and volume V are given by the formulae:

$$S \;=\; 2\pi \int_a^b dx\; y\sqrt{1 + (y')^2} \;. \tag{8.18}$$

$$V \;=\; \pi \int_a^b dx\; y^2 \;. \tag{8.19}$$

(a) Verify both formulae for the straight line $y = mx$ between $0 \leq x \leq 1$ (which generates a cone on revolution).
(b) Verify the volume formula for the semi-circle $x^2 + y^2 = 1$, $y > 0$, between $-1 \leq x \leq 1$.
(c) Calculate the surface area and volume of revolution for the curve $y = \sqrt{x}$ between $0 \leq x \leq 1$.
(d) Optional: Study the derivation of the formula.

19. If $y = xe^x$, find dy/dx. Hence, or otherwise, deduce $\int_0^1 xe^x \, dx$.

20. Prove that $\int_a^b f(x)dx = -\int_b^a f(x)dx$.

21. Evaluate $\dfrac{d}{dx}\ln(\sec x)$. Hence deduce the value of $\int_0^{\pi/4} \tan x \, dx$.

8.6 Challenges

1. Evaluate each of the following:
   (a) $\int_0^\infty x^n e^{-x} \, dx$ for $n > 0$ integral.
   (b) $\int_0^\infty e^{-x^2} \, dx$.
   (c) $\int_{-\infty}^\infty e^{-ax^2+bx+c} \, dx$ for $a > 0$.
   (d) $\int_{-\infty}^\infty x^n e^{-ax^2+bx+c} \, dx$ for $a > 0$ and $n > 0$ integral.

2. Using calculus, obtain an expression for the area of an ellipse, and in particular that of a circle.

3. Determine $\int \ln x \, dx$ as follows:
   (a) Sketch the curve $y = \ln x$ and evaluate explicitly $I_1 \equiv \int_a^b x \, dy$. (Note: $x = e^y$.)
   (b) Mark the area I_1 on your diagram for the curve.
   (c) Let $I_2 \equiv \int_{ea}^{eb} y \, dx$. Mark this area on the same diagram and deduce an expression for its area.
   (d) Keeping a as a constant but treating $b \rightarrow y$ as a variable,

deduce the explicit form for $\int \ln x \, dx$. Verify your result by differentiation.

8.7   Inquire and Investigate

1. (a) How did Archimedes and others in the ancient world determine areas and volumes of various bodies if integral calculus was supposedly only invented much later?
   (b) Review Archimedes' proof that the area of a circle is πR^2.

2. Is it possible for a curve $y = f(x)$ to have a definite integral, $\int_a^b f(x) \, dx$, even though at some points in the interval $[a, b]$ the function $f(x)$ does not have a well-defined derivative? Give explicit examples to support your discussion.

3. Is it possible for a three-dimensional figure to have infinite surface area but finite volume?

4. The method of "integration by parts" is a technique for evaluating some integrals. Study the method and apply it to compute some integrals.

Chapter 9

Supplementary Questions

"... encouraged me to compose a short work on calculating by al-jabr and al-muqabala, confining it to what is easiest and most useful in arithmetic".

Al-Khwarizmi

The questions in Sets A and B may be used for self-assessment. The answers are not provided; do discuss with your co-adventurers or check our website for potential hints. Sections (3) and (4) contain more questions of the type found at the end of each chapter: They go beyond the typical school syllabus for this level.

9.1 Set A

1. A farmer wins a land lottery which entitles him to select a piece of land under the following two conditions: (i) The shape must be that of a right-angled triangle with one perpendicular side 10 m long; (ii) The chosen site must be fenced using not more than 45 m of fencing material. Can you help the farmer select the largest possible area given the constraints?

2. (a) Solve for all the roots of the equation $x^3 - 2x^2 + 1 = 0$.
   (b) Hence solve the equation $(y+1)^3 - 2y^2 = 1 + 4y$.

3. With the aid of a sketch or otherwise, prove that the function $f(x) = \ln x - x + 1$ has only one real zero, at $x = 1$.

4. (a) Use calculus to find the slope of the tangent to the circle $x^2 + y^2 = 4$ at the point P whose coordinates satisfy $x = 1$ and $y > 0$.

   (b) Use trigonometry to find the slope of the radial line through P and hence the slope of the tangent at P. Compare your answer to that in part (a).

5. Show that in the binomial expansion of $\left(x^2 - \dfrac{1}{x}\right)^n$, with n a positive integer, there will be no constant term unless n is a multiple of three. Can the constant term be negative? Explain.

6. Prove the identity $\csc^2(x) + \sec^2(x) - \cot^2(x) = 2 + \tan^2(x)$.

7. Eona could not sleep well during an overnight camping trip with her fellow Girl Guides; whenever awake, she would measure the temperature of her surroundings. After her trip she decided to model the day's temperature variation approximately using the function $f(t) = A + B\sin(\omega t + C)$, with $f(t)$ representing the temperature in degrees Celsius, t hours after midnight.

   (a) Help Eona determine the constants A, B, ω, C using the following data she has collected: The maximum and minimum temperatures were respectively $24°C$ and $32°C$; the temperature at midnight was $28°C$; the temperature at $t = 6$ was $26°C$. (You may assume $0 < \omega < 3\pi/2$).

   (b) According to her model, what was the temperature at $t = 3$ and $t = 12$?

8. In the trapezium $ABCD$, DC is parallel to AB while the diagonals AC and BD intersect at E. The area of triangle ADC is 100 and $\angle CAB = 30°$. It is also known that $DC = 8$ and $EC = 4$. Find

   (a) The area of triangle ECB.

   (b) The area of triangle EAB.

9. Find the area bounded by the curves $y = e^x$, $y = 2$ and $x = 2$.

10. Newton's law of cooling describes how an object cools:
$$\frac{dT}{dt} \propto -k(T - T_a)$$
where $T(t)$ is the varying temperature of the

object, t is time, T_a is the constant ambient temperature, and $k > 0$ is a constant.

(a) Defining $y = T(t) - T_a$, show that $\dfrac{dy}{dt} = -ky$.

(b) Show that $y = Ae^{-kt}$ is a solution of the last equation.

(c) Hence obtain an explicit formula for $T(t)$ if $T(t = 0) \equiv T_i$.

(d) If $k = 0.06$ per minute, determine how long it would take a body at $100°$C to cool to $50°$C if $T_a = 30°$C.

11. Find the area bounded between the curves $x^2 + y^2 = 9$ and $y = 1$ for $y \geq 1$.

12. The cubic polynomial $f(x) = 3x^3 + 5x^2 - 6x - 8$ has one positive root which Neo wishes to find by a process of approximation.

(a) By looking at the turning points of $y = f(x)$ and sketching the curve, confirm that there is exactly one positive root.

(b) Verify that $f(1) < 0$ and $f(2) > 0$. Hence explain why the root must be between $x = 1$ and $x = 2$.

(c) Write the root as $x = 1 + z$ and substitute this into the original equation. Assuming z is small, obtain a linear equation for z by ignoring terms such as z^2 and higher orders. Solve the linear equation to obtain the first estimate for the root, x_1.

(d) Repeat part (c) but now keep the z^2 terms to obtain a quadratic equation; find a better estimate of the root.

(e) Instead of the procedure in part (d), try this alternative method after part (c): After obtaining x_1, write $x = x_1 + z$ and substitute it into the original equation. Keeping only linear terms, find an estimate for z and hence a second approximation for the root. (This will be the second iteration of the linear approximation).

(f) Compare your answers in parts (c), (d) and (e) with the exact answer which is $x = 4/3$.

(g) Optional: Can you use the approximation methods above to determine the other roots of the equation?

9.2   Set B

1. A quadratic polynomial in x leaves a remainder of 2^n when it is divided by $(x - n)$ for $n = 1, 2$ or 3.
   (a) Find the polynomial.
   (b) What remainder does the polynomial give when it is divided by $(x - 4)$?

2. The thin lens formula relates object distance u to image distance v and is given by $\dfrac{1}{u} + \dfrac{1}{v} = \dfrac{1}{f}$ where f is the focal length of the lens. For constant f, find a partial fraction expression for v in terms of u.

3. Find the shortest distance from the origin to the curve $xy = 1$ for $x > 0$.

4. In addition to gravitational acceleration, a particle experiences air-resistance as it falls along its vertical trajectory: Its net acceleration is given by $\dfrac{dv}{dt} = g - bv$, where $v(t)$ is the instantaneous velocity and b, g are positive constants (taking the downward direction as positive).
   (a) Show that $v(t) = g(1 - e^{-bt})/b$ satisfies the acceleration equation and the initial condition $v(0) = 0$.
   (b) Determine the constant velocity, known as "terminal velocity", which the particle will reach if it continues to fall for a long time.
   (c) Optional: Discuss one application of terminal velocity.

5. A student claims that the area of a particular triangle is given by the formula $A = \dfrac{a^2 \tan \alpha}{2}$ where a is the length of a short side and α an angle adjacent to it. Find all the angles of the triangle in terms of α and all its sides in terms of a and α.

6. A student comes across a trigonometric identity in an ancient manuscript: $\tan^2 A - trg^2 A = \tan^2 A \cdot trg^2 A$.
   Assuming that tan is the usual trigonometric function, help the student determine, in simplest terms, all possible trigonometric function(s) that "trg" might represent.

7. When a natural oscillator, such as a pendulum, experiences an external driving force which is periodic, its motion can become amplified, leading to the phenomenon of resonance. Consider the driven harmonic oscillator equation. The displacement $x(t)$ satisfies: $\ddot{x} + a^2 x = F \cos bt$, where a, F and b are constants.

   (a) Verify that when $F = 0$, the oscillator equation has the periodic solution $x(t) = A \cos at$ for some constant A.

   (b) For $F \neq 0$, assume a solution of the form $x(t) = B \cos bt$ and determine B explicitly.

   (c) Show that as the "driving frequency" b approaches the "natural frequency" a, the amplitude B increases.

   (d) Optional: In reality the amplitude cannot increase without bound. What limits its growth? What are some applications of resonance?

8. Find the area bounded between the curves $x^2 + y^2 = 9$, $y = 1$ and $y = 2$, for $2 \geq y \geq 1$.

9. According to quantum mechanics, which describes the property of matter at atomic scales, any "particle" also has wave characteristics. A particle confined to a one-dimensional box, $0 \leq x \leq L$, has an associated de Broglie-Schrodinger wave described by $y(x) = A \sin \dfrac{2\pi x}{\lambda}$ where A is the amplitude and λ the wavelength of the wave.

   (a) Using the condition $y(L) = 0$, find the possible values for λ.

   (b) Optional: How are the values of λ related to the quantised energies of the particle? Does this example have any technological applications?

10. (a) Prove that for m, n integers, the following orthogonality condition holds:

$$\int_{-\pi}^{\pi} \sin(mx) \, \sin(nx) \, dx = 0, \text{ if } m \neq n.$$

   (b) What happens when $m = n$?

   (c) Optional: What is the application of such relations in the field of Fourier Transforms?

11. Eona wishes to solve the equation $y(x) \equiv e^x + x - 2 = 0$ using the following approximation scheme:

    (a) Firstly, verify that $y(1) = e - 1 > 0$.

    (b) Construct an equation for the tangent line at $x = 1$. Sketch the curve and the tangent line.

    (c) Use the tangent line to approximate (that is, replace) the curve near $x = 1$. Estimate the solution of $y(x) = 0$ using the equation for the tangent line instead of the curve.

    (d) Check the accuracy of your answer to part (c).

    (e) Use the approximate solution in part (c) as the starting point for a new tangent line and repeat the process to obtain another approximation for the root.

    (f) Optional: The method described can be generalised and used in a computer algorithm. Read about "root finding" algorithms such as Newton's method.

9.3   Additional Challenges

These questions typically go beyond the usual syllabus. Attempt as many of the questions within 3 hours.

1. Solve for real x: $\sqrt{3x - 1} - \sqrt{x - 2} + \sqrt{2x - 6} - \sqrt{4x - 5} = 0$.

2. A triangle is inscribed inside a unit square such that one of the vertices coincides with a vertex of the square. Determine, with proof, the triangle of maximum area that can be so inscribed.

3. Solve for x: $(x - 2\sqrt{x}) + 4(\dfrac{1}{x} - \dfrac{1}{\sqrt{x}}) + 1 = 0$.

4. The 200 students of a school are teamed up into groups of either 3 or k. If the total number of teams formed is 68, determine the possible values of k.

5. In addition to gravitational acceleration, a particle experiences air-drag as it moves along its vertical trajectory: Its acceleration is given by $a = -g - bv$ where v is the instantaneous velocity and b, g are positive constants. If its velocity v and displacement y

at $t = 0$ are $v(0) = u > 0$ and $y(0) = 0$, determine

(a) $v(t)$.

(b) $y(t)$.

(c) The time when the particle reaches its maximum height.

(d) The maximum height reached by the particle.

(e) For b small, estimate the time when the particle returns to $y = 0$ (include the leading order correction in b.)

(f) For b small, estimate the velocity of the particle when it returns to $y = 0$.

6. Find the smallest $x > 0$ which solves:
$$5(\cos(x/2) - \sin(x/2)) = 4\cos x \,(\cos(x/2) + \sin(x/2)).$$

7. Suppose that $e^x \equiv \sum_{n=0}^{\infty} a_n x^n$ is the infinite series representation of the exponential function e^x, where a_n are constants to be determined.

(a) By repeated differentiation of both sides of the equation, and using $e^0 = 1$, determine the a_n and hence the explicit series for e^x.

(b) Similarly, obtain the infinite series for $\sin x$ and $\cos x$.

(c) Hence deduce Euler's formula $e^{ix} = \cos x + i\sin x$ where $i = \sqrt{-1}$.

(d) By squaring Euler's formula, derive the double angle formulae for $\sin 2x$ and $\cos 2x$.

9.4 Additional Investigations

1. Look up the defining equation for the spiral of Theodorus and investigate its unique mathematical properties. Are there naturally occurring phenomena described by such a spiral? How does it relate to the Archimedean spiral?

2. Study the history of the decimal system in use today. What impact did it have on the progress of mathematics?

3. Matrix multiplication is non-commutative, that is $AB \neq BA$ in general. Such non-commutative quantities play a pivotal role

in quantum mechanics. Read about Heisenberg's formulation of Matrix Mechanics.

4. Fermat's Last Theorem, an example of a Diophantine problem, was only proven in the 1990's by A. Wiles despite having been conjectured a few hundred years earlier. Read about its history and impact.

5. Several quotations attributed to famous mathematicians appear in this book. Choose some of them and investigate the context in which they occurred.

6. Choose one of the mathematicians listed below, or another of your choice, and write an essay on her/his life and times, and main contribution to mathematics.
Hypatia, Archimedes, Aryabhata, Bhaskara, Omar Khayyam, Al-Khwarizmi, Sophie Germain, Emmy Noether, Newton, Emilie du Chatelet, Liu Hui, Lagrange, Descartes, Legendre, Leibniz, Diophantine, Fermat, Bernoulli, Pascal.

7. The cables supporting the deck of a suspension bridge do not take the shape of a catenary but rather resemble more closely a parabola. Why is that so? To what extent do the catenary and parabola differ?

8. In Euclidean space, the distance between two points as calculated using Pythagoras' formula is an invariant, that is, it is independent of the coordinate system used.
In Einstein's Special Theory of Relativity, space and time are united into a Minkowskian spacetime which also has an invariant measure (that is, "distance" measure) between two points. However in Minkowskian spacetime, the invariant measure can be zero or negative. Investigate this new spacetime and the physical significance of its invariant measure.

Chapter 10

Escapades

"This leads me to confess that
I am not as completely unknown to you as you might believe,
but that fearing the ridicule attached to a female scientist,
I have previously taken the name of M. LeBlanc
in communicating to you those notes..."
Sophie Germain

10.1   Puzzles

Try your hand at these classic puzzles, using any method that comes to mind. Some of the puzzles have a rich mathematical framework associated with them which you can explore further in a self-study project.

1. **Three boats** are heading down the river when the first boat, A, starts to sink. To stabilise the situation, some passengers are transferred to boats B and C so that the number of passengers in each of B and C is doubled. Some time later it is boat B that experiences difficulty and so again a transfer of passengers is done from B to A and C so that the latter two boats each have their numbers doubled. During the final stretch of the trip it is boat C that has problems and has to transfer some passengers to A and B; again, after the transfer the number of passengers in A and B has doubled. When the three boats reach their destination, it

is found that they each carry the same number of passengers!
(a) What is the smallest number of passengers each boat started
with?
(b) Find all solutions for which the total number of passengers
in the three boats is less than 100.
(c) There is more than one way to solve this problem. See if you
can find an efficient way.

2. **Tower of Hanoi.** n discs, each of a different diameter, are
stacked one on top of the other in order of size, with the largest
at the bottom. The initial stack lies over a square labelled A.
You are asked to move all the discs over to another square B
with the following rules:
(i) You may only move one disc at a time, the top-most in any
pile.
(ii) A disc may not be placed over a smaller disc.
(iii) You may use an intermediate square, marked C, to place
your discs, subject to the same rules (i) and (ii).

(a) Find the smallest number of moves, S_n, that accomplishes
the task for $n = 2, 3, 4$.
(b) Show that the following recursive relation applies for any
$n \geq 1$, $S_{n+1} = 2S_n + 1$.
(c) Hence, derive an explicit expression for S_n, or guess a form
for S_n and show that it satisfies the recursive relation.

3. **Rupert's Cubes.** You have a wooden cube with unit sides,
and another iron cube. What is the largest size of the iron cube
that can pass through a hole made in the wooden cube, without
destroying the integrity of the wooden cube? (HINT: It might
help to first find the largest square that lies within a unit cube.)

4. **Three Smart Students** are blind-folded while a small white
hat is placed on each of their heads. They are told that not
all the hats are black. With the blind-folds removed, each is
asked to deduce whether the colour of their own hat is white
or black (obviously without removing or looking at their own
hat). Almost simultaneously, all three students shout out the

right answer. How did they figure it out? (HINT: Each student knows that the other students are equally smart).

5. **The seven bridges of Konigsberg** connected the parts of the city which were separated by rivers, as shown in Fig.(10.1). The inhabitants wondered if it was possible to walk a path that crossed each bridge exactly once. What do you think?

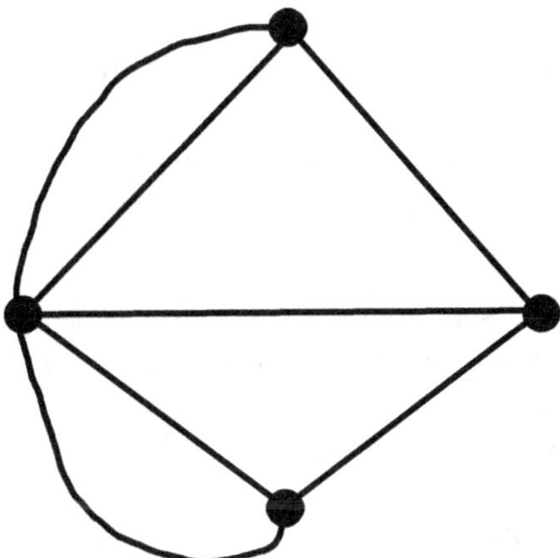

Figure 10.1: A sketch (graph) showing the seven bridges as edges. The four vertices are the land masses.

6. **You are as tall as Everest**. Let x be your height and y that of Mount Everest. The average height is $z = (x + y)/2$: From this last equation you can derive two relations, $x = 2z - y$ and $x - 2z = -y$. Multiply those two relations to get $x^2 - 2xz = y^2 - 2zy$, and then complete the square on both sides by adding z^2. Hence you have $(x - z)^2 = (y - z)^2$; take the square root of both sides to get $x - z = y - z$, hence $x = y$!
You see it, but you do not believe it, do you? What went wrong?

10.2 Wonderful Theorems

There are several beautiful and wonderful theorems in mathematics. Here we list a few of our favourites:

1. **Fundamental Theorem of Algebra:** Every polynomial of degree n has n roots within the field of complex numbers (repeated roots are counted with their multiplicity). The statement includes polynomials with complex numbered coefficients, for example $x^2 - (1 + 2i)x + 1 = 0$ where $i = \sqrt{-1}$.

   Historically, solving equations such as $x+1 = 0$ and $x^2+1 = 0$ required us to first go beyond positive numbers to include negative, and then imaginary, and hence complex numbers. Essentially, the theorem says that we do not have to further enlarge the set of numbers: All polynomial equations can be solved within the complex number system.

   The theorem was first proven by Gauss but there was a subtle gap in his original proof. A rigorous proof was given by Argand.

2. **Euler's Formula:** $e^{ix} = \cos x + i \sin x$
   This remarkable formula, which relates the trigonometric functions to the exponential functions using complex numbers, is not only considered "beautiful" by many mathematics enthusiasts, it is also extremely important in the mathematical sciences and engineering. The special case $e^{i\pi} = -1$ is particularly striking.

   De Moivre's formula, $(\cos x + i \sin x)^n = \cos(nx) + i \sin(nx)$ is an easy consequence of Euler's formula. See challenge question (7) in Chapter (9).

3. **Cantor's Theorems.** Cantor showed that the set of rational numbers, though infinite, is "countable": It is precisely as large as the infinite set of natural numbers. He further showed that the set of real numbers is uncountable, so there are more real numbers than natural numbers.

   In other words, the infinity of real numbers is larger than the infinity of natural numbers! One consequence is that there are more irrational numbers than rational numbers.

Cantor proved more: He showed that the set of **algebraic numbers** is countable; "algebraic numbers" are those that are the roots of a polynomial with integral coefficients, for example, all rational numbers and some irrationals such as $\sqrt{2}$ are algebraic. Thus the sub-set of real numbers which are not algebraic, called "transcendental", is larger than the sub-set of algebraic numbers! Examples of **transcendental numbers** are π and e.

4. **Descartes-Euler's Polyhedral Formula.** Consider a convex polyhedron with V vertices, E edges and F faces. The formula states that for convex polyhedra $\chi \equiv V + F - E = 2$. For example, a cube has $V = 8$, $F = 6$, $E = 12$ while a tetrahedron has $V = 4$, $F = 4$, $E = 6$. The formula also applies to planar graphs.

   The **Euler characteristic**, denoted by χ (pronounced "kai"), describes the "topological" characteristics of a surface. That is, properties that do not change when the surface is bent. A surface (imagine it to be made of rubber) which can be bent into the shape of a sphere, without tearing, will have $\chi = 2$.

   On the other hand, a rubber donut cannot be stretched to form a sphere, and vice-versa, because of the "hole" in the former. This is stated by saying that the donut is "topologically inequivalent" to a sphere. In fact the donut has $\chi = 0$.

10.3   Unsolved Problems

Here is a small list of the many problems in mathematics which still remain unsolved despite the efforts of numerous professional and amateur mathematicians. For some, partial results have been obtained. Try to understand the problems by experimenting with some special cases; see the respective write-up in Wikipedia for more details if you are interested.

1. **Collatz Conjecture.** Start with any natural number n. If it is odd, multiply it by 3 and add 1 to get $3n + 1$; but if n is even, divide it by 2 to obtain $n/2$. Keep repeating the process on each result to generate a sequence of numbers. For example:

7, 22, 11, 34, 17, 52, 26, 13, 40, 20, 10, 5, 16, 8, 4, 2, 1, ...
Conjecture: Regardless of the starting value n, the sequence will always reach 1 (after which it will enter into a cycle $1 \to 4 \to 2 \to 1$).
Test the conjecture for a few numbers to convince yourself that it is plausible. Using computers, the conjecture has been verified up to very large numbers, but that is not proof that it is true for all n.

2. **The Perfect Cuboid**: *Does there exist a rectangular prism with integer sides, integer face diagonals and an integer body diagonal?* If the body diagonal is not required to be an integer then one has an "Euler brick", several solutions of which are known, for example a brick with sides (44, 117, 240). However, to date, it is not known if a perfect cuboid exists (extensive computer searches have not yielded any).

3. **Catalan's Constant** is defined by the infinite series
$G = \frac{1}{1^2} - \frac{1}{3^2} + \frac{1}{5^2} - \frac{1}{7^2} + \frac{1}{9^2} - \frac{1}{11^2} + \ldots = 0.915965594177\ldots$
Is G rational or irrational? Though most people believe it is irrational, a proof is lacking. As a comparison, it is known that the following series with only positive signs:
$\frac{1}{1^2} + \frac{1}{3^2} + \frac{1}{5^2} + \frac{1}{7^2} + \frac{1}{9^2}\ldots = \pi^2/8$ is irrational because π^2 is known to be irrational.

4. **Lonely Runner Conjecture**: At time $t = 0$, k runners are at the same position on a circular track of unit length. Each athlete has a distinct running speed. A runner becomes "lonely" if he is at distance of at least $1/k$ from any other runner.
Conjecture: Every runner will get lonely at some time.
Try to understand the problem first for $k = 1, 2, 3$. At time of writing, proofs in the affirmative exist only for individual cases up to $k = 7$.

5. **Odd and Perfect?** A "perfect number" is defined as a positive integer which equals the sum of its proper positive divisors; for example, $6 = 1 + 2 + 3$ and $28 = 1 + 2 + 4 + 7 + 14$ are perfect. All known perfect numbers are even. *Are there any odd ones?*

10.3.1 Prizes for Mathematics

There are several prizes which recognise the contribution of individuals to the development of mathematics, the most prestigious probably being the Field's medal and the Abel Prize.

There are also prizes which are offered for the solution of specific mathematical problems. The ones with the largest monetary value are the seven Millennium Prize Problems, each of which can earn the solver US \$ 1,000,000. The most well-known of the seven problems are the "P vs NP" problem (related to the Travelling Salesman problem) and the Riemann Hypothesis.

Two of the Millennium problems involve questions in theoretical physics, related to the Yang-Mills equations and the Navier-Stokes equations.

One of the seven problems, the Poincare Conjecture, was solved in 2003 by G. Perelman. However, for personal reasons, Perelman declined the million dollar prize (and also reportedly the Fields medal nomination).

Most mathematicians, like most scientists, do what they do because they enjoy it. Being human, they do appreciate recognition of their work through various awards. The monetary prizes are rare, and are meant mainly to provide them with financial independence to continue their passion.

Walk
Are you making no progress on a problem despite your best attempts? Take a break, go for a walk, or engage in some other activity which you enjoy. After an initial conscious concentration on some problem, the sub-conscious reflection during a relaxing activity might later produce some useful insights into the problem.

Appendix A

Formulae and Summary

Note to student: It is not useful to memorise all the formulae, partly because many of the complicated formulae may be obtained from the simpler ones. Rather, you should familiarise yourself with the simpler formulae by solving a number of practice questions which use those. Consult your school to check which formulae you are expected to know by heart for examinations, and which others will be provided to you in a help sheet.

Exponents and Logarithms

- For $a, b > 0$,

$$a^{-1} = \frac{1}{a}. \tag{A.1}$$

$$a^0 = 1. \tag{A.2}$$

$$a^{xy} = (a^x)^y. \tag{A.3}$$

$$a^x a^y = a^{x+y}. \tag{A.4}$$

$$(ab)^x = a^x b^x. \tag{A.5}$$

$$\sqrt{ab} = \sqrt{a}\,\sqrt{b}. \tag{A.6}$$

- For $a > 1$, $y = a^x$ is positive and an increasing function of x along the real line; for $0 < a < 1$, a^x is a decreasing function of x.

- The basic relation between exponents and logarithms is

$$y = b^x \Leftrightarrow x = \log_b y \, . \tag{A.7}$$

- For $a, b > 0$, $a \neq 1$, $b \neq 1$ and $P, Q > 0$,

$$\log_b 1 \;=\; 0. \tag{A.8}$$
$$\log_b PQ \;=\; \log_b P + \log_b Q \, . \tag{A.9}$$
$$\log_b \frac{P}{Q} \;=\; \log_b P - \log_b Q \, . \tag{A.10}$$
$$\log_b P^c \;=\; c \log_b P \, . \tag{A.11}$$
$$\log_b P \;=\; \frac{\log_a P}{\log_a b} \, . \tag{A.12}$$
$$\log_b a \;=\; \frac{1}{\log_a b} \, . \tag{A.13}$$

- For $a > 1$ and $x > 0$, $y = \log_a x$ is an increasing function of x; it is positive for $x > 1$, negative for $x < 1$ and vanishes at $x = 1$. The graph of $y = \log_a x$ may be obtained by reflecting the exponential curve $y = a^x$ about the line $y = x$.

Binomial Expansions

For a positive integer n, the Binomial Theorem states

$$(a+b)^n \;=\; a^n + \binom{n}{1} a^{n-1}b + \binom{n}{2} a^{n-2}b^2 + \binom{n}{3} a^{n-3}b^3 + \ldots$$
$$+ \binom{n}{r} a^{n-r}b^r + \ldots + \binom{n}{n-1} ab^{n-1} + b^n \tag{A.14}$$

where the binomial coefficient is defined by

$$^nC_r \equiv \binom{n}{r} = \frac{n!}{r!(n-r)!} \tag{A.15}$$

and $n! \equiv n \times (n-1) \times (n-2) \times \ldots \times 2 \times 1$ with $0! \equiv 1$.

Note that the $(r+1)$-th term in the above expansion is given by $\binom{n}{r} a^{n-r}b^r$.

Polynomials

- **Remainder Theorem**: If the polynomial $P(x)$ is divided by $(x - \alpha)$, the remainder is $P(\alpha)$.
  In other words, $P(x) \equiv (x - \alpha)Q(x) + R$ with $P(\alpha) = R$.

- **Factorisation Theorem**:
  $P(\alpha) = 0$ if and only if $P(x) \equiv (x - \alpha)Q(x)$.

- For the quadratic equation $ax^2 + bx + c = 0$,

$$x = \frac{-b \pm \sqrt{\Delta}}{2a}. \tag{A.16}$$

where

$$\Delta \equiv b^2 - 4ac \tag{A.17}$$

is called the **discriminant**. The two roots are real if and only if $\Delta \geq 0$; the case $\Delta = 0$ corresponds to a repeated root. Denoting the two roots of a quadratic equation by α and β, then

$$\begin{aligned} \alpha + \beta &= \frac{-b}{a}. \\ \alpha\beta &= \frac{c}{a}. \end{aligned} \tag{A.18}$$

Other common relations are

$$\begin{aligned} \alpha - \beta &= \pm\sqrt{(\alpha - \beta)^2} \\ &= \pm\sqrt{(\alpha + \beta)^2 - 4\alpha\beta}. \end{aligned} \tag{A.19}$$

and

$$\alpha^2 + \beta^2 = (\alpha + \beta)^2 - 2\alpha\beta. \tag{A.20}$$

- The shape of the curve $y = ax^2 + bx + c$ is determined by the sign of a: When $a > 0$, the curve has a minimum point while for $a < 0$ it has a maximum. For a cubic curve $y = ax^3 + bx^2 + cx + d$, the sign of the leading coefficient again determines its main shape. If $a > 0$, the curve rises upwards for large positive x and decreases for large negative x. In between, it might have a local minimum and a local maximum.

Partial Fractions

- Given a rational function $P(x)/Q(x)$, its decomposition into partial fractions is as follows: First use long division to reduce the degree of the numerator to below that of the denominator. Next, each factor of $(x - a)$ in $Q(x)$ would require a partial fraction $A/(x - a)$.

- If the factor is repeated in Q, for example $(x-a)^2$, then one uses two partial fractions $A_1/(x - a)$ and $A_2/(x - a)^2$ for that factor.

- If Q contains a term that cannot be factorised (using real numbers), for example $x^2 + x + 1$, then the partial fraction for that term is of the form $(Ax + B)/(x^2 + x + 1)$, that is, the numerator is one degree lower than the denominator.

Simultaneous Equations

- A $m \times n$ matrix A multiplying from the left a $n \times p$ matrix B yields a $m \times p$ matrix C; that is $C = AB$. The c_{ij} element of C is obtained by multiplying the i-th row of A to the j-th column of B term by term and adding the pieces.

- The pair of simultaneous equations in the variables (x, y):

$$ax + by \;=\; e \tag{A.21}$$
$$cx + dy \;=\; f, \tag{A.22}$$

  may be written as the matrix equation $MX = R$ with

$$M = \begin{pmatrix} a & b \\ c & d \end{pmatrix}, \tag{A.23}$$

$$X = \begin{pmatrix} x \\ y \end{pmatrix} \tag{A.24}$$

  and

$$R = \begin{pmatrix} e \\ f \end{pmatrix}. \tag{A.25}$$

- The **determinant** of the 2×2 matrix M is defined by $\det(M) = ad - bc$. If $\det(M) \neq 0$ then one may define an **inverse matrix**

$$M^{-1} = \frac{1}{\det(M)} \begin{pmatrix} d & -b \\ -c & a \end{pmatrix} \qquad (A.26)$$

and the solution to the problem is

$$X = M^{-1}R = \frac{1}{\det(M)} \begin{pmatrix} de - bf \\ -ce + af \end{pmatrix}. \qquad (A.27)$$

- If $\det(M) = 0$ the matrix is termed **singular** and there is no inverse matrix. The inverse matrix satisfies

$$MM^{-1} = M^{-1}M = \begin{pmatrix} 1 & 0 \\ 0 & 1 \end{pmatrix}, \qquad (A.28)$$

the last matrix being the **identity matrix**, usually denoted by the letter I.

Trigonometry

- For angle measurements in **radian**, $\theta \equiv s/R$ where s is the arc length of circle subtended by that angle. Therefore 2π radians equals $360°$.

- For a right-angled triangle ABC, with $C = 90°$, $\sin A = a/c$, $\cos A = b/c$ and $\tan A = \sin A / \cos A = a/b$, where the small case letters denote lengths opposite the corresponding angles.

- The sin and cos functions range over the interval $[-1, 1]$ while the tan functions ranges over the real line.

- Periodicity: $\sin(\theta + 360°) = \sin(\theta)$, $\cos(\theta + 360°) = \cos(\theta)$ and $\tan(\theta + 180°) = \tan(\theta)$.

- The **principal values** for the inverse functions $\sin^{-1}$ and $\tan^{-1}$ are those that lie in the range $-\pi/2 \leq y \leq \pi/2$ while for the $\cos^{-1}$ function has the range $0 \leq y \leq \pi$.

- The functions csc, sec and cot are reciprocals of the sin, cos and tan functions ($\csc x$ is also written as $\operatorname{cosec} x$).

- Phenomena that are described by an equation of the form $y(x) = A + B\sin(kx + C)$ are called **sinusoidal**.

- Some identities:

$$\sin^2 A + \cos^2 A \;=\; 1 \;. \tag{A.29}$$
$$1 + \tan^2 A \;=\; \sec^2 A \;. \tag{A.30}$$
$$\sin(-A) \;=\; -\sin A \;. \tag{A.31}$$
$$\cos(-A) \;=\; \cos A \;. \tag{A.32}$$
$$\tan(-A) \;=\; -\tan A \;. \tag{A.33}$$

- Addition Formulae:

$$\sin(A \pm B) \;=\; \sin A \, \cos B \pm \cos A \, \sin B \;. \tag{A.34}$$
$$\cos(C \pm D) \;=\; \cos C \, \cos D \mp \sin C \, \sin D \;. \tag{A.35}$$
$$\tan(A \pm B) \;=\; \frac{\tan A \pm \tan B}{1 + \tan A \, \tan B} \;. \tag{A.36}$$

- Special Cases of addition formulae:

$$\sin(90^\circ \pm A) \;=\; \cos A \;. \tag{A.37}$$
$$\sin(180^\circ \pm A) \;=\; \mp \sin A \;. \tag{A.38}$$
$$\cos(90^\circ \pm A) \;=\; \mp \sin A \;. \tag{A.39}$$
$$\cos(180^\circ \pm A) \;=\; -\cos A \;. \tag{A.40}$$
$$\tan(90^\circ \pm A) \;=\; \mp \cot A. \tag{A.41}$$

- Double Angle Formulae:

$$\sin 2A \;=\; 2\sin A \cos A \;. \tag{A.42}$$
$$\cos 2A \;=\; 2\cos^2 A - 1 = 1 - 2\sin^2 A = \cos^2 A - \sin^2 A \;. \tag{A.43}$$
$$\tan 2A \;=\; \frac{2\tan A}{1 - \tan^2 A} \;. \tag{A.44}$$

- Factor Formulae:

$$\sin A + \sin B = 2\sin\frac{A+B}{2}\cos\frac{A-B}{2} . \qquad (A.45)$$

$$\sin A - \sin B = 2\cos\frac{A+B}{2}\sin\frac{A-B}{2} . \qquad (A.46)$$

$$\cos A + \cos B = 2\cos\frac{A+B}{2}\cos\frac{A-B}{2} . \qquad (A.47)$$

$$\cos A - \cos B = -2\sin\frac{A+B}{2}\sin\frac{A-B}{2} . \qquad (A.48)$$

- R-Formulae:

$$a\cos\theta \pm b\sin\theta = R\cos(\theta \mp \alpha) . \qquad (A.49)$$

$$a\sin\theta \pm b\cos\theta = R\sin(\theta \pm \alpha) . \qquad (A.50)$$

with $R = \sqrt{a^2 + b^2}$, $\tan\alpha = b/a$.

Properties of Triangles

- For a right-angled triangle, a mnemonic is "**soh-cah-toa**", which summarises
$\sin = Oppositeside/Hypotenuse$,
$\cos = Adjacentside/Hypotenuse$ and
$\tan = Opposite/Adjacent$.

- Area of a triangle:

$$A = \frac{1}{2}ab \sin C . \qquad (A.51)$$

- The **sine rule**:

$$\frac{c}{\sin C} = \frac{b}{\sin B} = \frac{a}{\sin A} . \qquad (A.52)$$

Each ratio in the formula equals the diameter of a circle which circumscribes the triangle (the centre of the circle being at the intersection of the perpendicular bisectors of the three sides).

- The **cosine rule**:

$$c^2 = a^2 + b^2 - 2ab \cos C . \qquad (A.53)$$

Elevation, Depression and Bearing

- If a point P is above the horizontal through the observation point O, then the angle that OP makes with the horizontal is the **angle of elevation** of P. Similarly if P were below the horizontal then the corresponding angle would be termed **angle of depression**.

- A **bearing** denotes a direction relative to North. It is usually expressed by a clockwise angle measured in degrees. For example, 065° would refer to the direction 65° clockwise from North. Such bearings are called "absolute bearings".

- It is also convenient to use "relative bearings", which define an angle relative to a chosen axis, for example the axis along which an aircraft is pointing.

Coordinate Geometry

- If (x, y) and (x_1, y_1) are two points on the line,

$$\frac{y - y_1}{x - x_1} = m = \tan \alpha \,, \tag{A.54}$$

where $0 \le \alpha < \pi$ is the angle the line makes with the positive x-axis. The equation for a straight line may be written in the form

$$y = mx + c \tag{A.55}$$

where m is the slope and c the intercept on the y-axis.

- If another line is perpendicular to (A.55), its slope must be $-1/m$.

- The mid-point of a line joining two points (x_1, y_1) and (x_2, y_2) is given by $\left(\dfrac{x_1 + x_2}{2}, \dfrac{y_1 + y_2}{2} \right)$.

- The equation for a circle is

$$(x - x_0)^2 + (y - y_0)^2 = r^2 \,, \tag{A.56}$$

where (x_0, y_0) is the centre and r the radius. Given the form

$$x^2 + y^2 - 2ax - 2by + c = 0, \qquad \text{(A.57)}$$

for some constants a, b, c, you can complete the squares to get $(x-a)^2 + (y-b)^2 = a^2 + b^2 - c$; if $c < a^2 + b^2$ then (5.4) represents a circle with centre (a, b) and radius $R = \sqrt{a^2 + b^2 - c}$.

- A tangent to a circle at any point P is perpendicular to the line OP where O is the centre of the circle. Therefore the **normal** to the circle at P lies along OP.

- Given three points $A(x_1, y_1)$, $B(x_2, y_2)$ and $C(x_3, y_3)$, with their relative order being anti-clockwise, the area of the triangle ABC may be obtained from the formula

$$A = \frac{1}{2} \begin{vmatrix} x_1 & x_2 & x_3 & x_1 \\ y_1 & y_2 & y_3 & y_1 \end{vmatrix} \qquad \text{(A.58)}$$

where the terms are generated as follows. First, start at the left of the top row and multiply each term in the top row by a term one step to the right in the bottom row, adding the pieces: $x_1 y_2 + x_2 y_3 + x_3 y_1$. Then start at the right of the top row and multiply each term in the top row by a term one step to the left in the bottom row, adding the pieces: $x_1 y_3 + x_3 y_2 + x_2 y_1$. Finally, subtract the two contributions and include the overall $1/2$ to get

$$A = \frac{1}{2}(x_1 y_2 + x_2 y_3 + x_3 y_1 - x_1 y_3 - x_3 y_2 - x_2 y_1). \qquad \text{(A.59)}$$

- The area of a polygon may be determined by dividing the polygon into triangles and using the above formula for each triangle.

Plane Geometry

- A **cyclic polygon** is one whose vertices lie on the circumference of a circle. Every triangle is in fact cyclic.

- **Inscribed Angle Theorem**
  Let A and B be two points on the circumference of a circle. The angle subtended by A and B at the centre of the circle is twice that subtended at a point C on the circumference.

- **Tangent-Chord Theorem (Alternate Segment Theorem)**
  Let the triangle ABC be inscribed in a circle and a tangent drawn at point A. Let D be another point on the tangent line such that D and C are on opposite sides of the line AB. Then $\angle DAB = \angle BCA$.

- **Intersecting Chords Theorem**
  Let A, B, C, and D be points on the circumference of a circle and X the point of intersection of the lines AC and BD. Then the triangle ABX is similar to triangle DCX.

- **Tangent-Secant Theorem**
  Let A, B and C be points on the circumference of a circle and let the tangent at A meet the line CB produced at D. Then $(DA)^2 = DB \times DC$.

Differential Calculus

- Let f and g be functions of x, and A, B, n represent constants in the following.

$$\frac{d}{dx} A x^n = A n x^{n-1}. \tag{A.60}$$

$$\frac{d}{dx} e^x = e^x. \tag{A.61}$$

$$\frac{d \ln x}{dx} = \frac{1}{x}. \tag{A.62}$$

$$\frac{d}{dx} \sin x = \cos x. \tag{A.63}$$

$$\frac{d}{dx} \cos x = -\sin x. \tag{A.64}$$

$$\frac{d}{dx} \tan x = \sec^2 x. \tag{A.65}$$

$$\frac{d}{dx}(f+g) \;=\; \frac{df}{dx} + \frac{dg}{dx}\,. \tag{A.66}$$

$$\frac{d(fg)}{dx} \;=\; f\frac{dg}{dx} + g\frac{df}{dx}\,. \quad \textbf{Product Rule} \tag{A.67}$$

$$\frac{d}{dx}\frac{f}{g} \;=\; \frac{gf' - fg'}{g^2}\,. \quad \textbf{Quotient Rule} \tag{A.68}$$

$$\frac{df(g(x))}{dx} \;=\; \frac{df}{dg} \times \frac{dg}{dx}\,. \quad \textbf{Chain Rule} \tag{A.69}$$

$$\frac{df}{dx} \;=\; 1/\left(\frac{dx}{df}\right)\,. \tag{A.70}$$

- The **stationary** points of $y = f(x)$ are those for which $\dfrac{df}{dx} = 0$. The stationary points are **local minima, local maxima, or points of inflexion.** If $\dfrac{d^2 f}{dx^2} > \;(<)\; 0$ at the stationary point, it is a local minimum (maximum). Stationary points with $\dfrac{d^2 f}{dx^2} = 0$ can be examined by checking the sign of $f'(x)$ on both sides of the point. Local extrema need not be global extrema.

- If Δx is small but not strictly zero, we may form the approximation

$$\Delta y \approx \left(\frac{dy}{dx}\right) \times \Delta x\,. \tag{A.71}$$

Integral Calculus

- **Fundamental Theorem of Calculus:**

$$\int_a^b f(x)dx = F(b) - F(a)\,, \tag{A.72}$$

where $F(x)$ is a function that satisfies

$$\frac{dF(x)}{dx} = f(x)\,. \tag{A.73}$$

Hence

$$\int_a^b \frac{dF(x)}{dx}\,dx = F(b) - F(a)\,. \tag{A.74}$$

- Indefinite integrals:

$$\int f(x)dx = F(x) + C\,,\qquad (A.75)$$

where (A.73) still holds and C is a constant of integration which can be fixed once we have more information about the problem.

- In the formulae for indefinite integrals below, f and g are functions of x while a, b, n are constants; C is a constant of integration.

$$\int (a + bx)^n\, dx = \frac{(a + bx)^{n+1}}{b(n + 1)} + C,\quad n \neq -1\,. \qquad (A.76)$$

$$\int \frac{1}{a + bx}\, dx = \frac{1}{b}\ln(a + bx) + C\,. \qquad (A.77)$$

$$\int \sin(a + bx)\, dx = \frac{-1}{b}\cos(a + bx) + C\,. \qquad (A.78)$$

$$\int \cos(a + bx)\, dx = \frac{1}{b}\sin(a + bx) + C\,. \qquad (A.79)$$

$$\int e^{(a+bx)}\, dx = \frac{1}{b}e^{(a+bx)} + C\,. \qquad (A.80)$$

$$\int (f + g)\, dx = \int f\, dx + \int g\, dx\,. \qquad (A.81)$$

- The following identity holds:

$$\int_a^c f(x)dx = \int_a^b f(x)dx + \int_b^c f(x)dx\,. \qquad (A.82)$$

- For calculating areas between the curve $y = f(x)$ and the x-axis, note that if $f(x) < 0$ within a region $x_1 \leq x \leq x_2$, the integral in that region would give a negative value and the area is then the negative of the integral.

- One may also evaluate the area between a curve and the y-axis. In this case the integral would be $\int_{y_1}^{y_2} x\, dy$.

Appendix B

Hints for Selected Exercises

Exponents and Logarithms

3. (a) Use eq.(1.1).

4. (a) Put $N(t) = N_0/2$ and use eq.(1.1).

4. (b) Use part (a) to get λ for this element, then use the decay law.

4. (c) Calculate λ for this substance using the decay law, then use the formula for half-life from part (a).

5. (b) Under what condition is the square root real?

5. (c) Write the equation as $E\sqrt{1 - v^2/c^2} = mc^2$ before taking $v \to c$.

6. (a) When λ is large, $1/\lambda$ is small and the right hand side of the formula must be small. So n_2 should be as close as possible to $n_1 = 1$.

7. (a) First compare the two terms with base 5, and separately the two terms with base 7, then arrange them.

7. (b) Every term here may be written as $a \log_2 3$ where a varies.

8. (a) Note that $8 = 2^3$ and $4 = 2^2$.

9. (c) Write $\log_5 27 = 3\log_5 3^3 = 3\log_5 3$ then use (1.10).

11. (a) $1/25 = 5^{-2}$.

11. (b) Express all terms as powers of 2 and 3 which are prime numbers. That is, note that $2^a = 3^b$ implies $a = b = 0$.

11. (d) Combine all the log terms into one piece and then solve the equation.

11. (e) Use (1.10) to write everything in terms of $\log_2$ and solve.

11. (f) Square both sides.

11. (g) Take the log of both sides.

12. (b) It might be easier to first write it as $2^6(1 + 0.01)^6$.

13. (c) It might be easier to first write $\left(x + \dfrac{1}{x}\right)^8$ as $\dfrac{(1 + x^2)^8}{x^8}$.

Polynomials and Rational Functions

1. (a) Set $a/b = x$; then solve the quadratic equation for x.

3. (a) Set $\sqrt{x} = u$.

3. (b) Set $\sqrt{x} = u$ and multiply the equation by u^2.

9. (c) Set $a \cdot (x^2 - 2x - 2) = x - 3a$ and solve for x, making the discriminant vanish to get two equal roots.

Simultaneous Equations

1. Let the number of passengers be A and B respectively and write each piece of information as an equation. For example, the first is $A + 1 = 2(B - 1)$.

2. Form the two conditions $3x + 5y = 200$, $x + y < 50$ and combine them into one inequality.

7. (b) Treat it as the two inequalities. $x - 1 > 3x - 4$ and $3x - 4 > -2 - x$.

Trigonometry

1. (a) Use the tan function.

3. (a) Write $t = x/(U \cos \theta)$ and substitute into the equation for $y(t)$.

3. (b) Set $y = 0$ and solve for $y = R$.

7. (a) Note $\cos 35° = \sin(90° - 35°) = \sin 55°$ and $\tan x = \sin x / \cos x$.

9. (a) Use $\sin^2 x + \cos^2 x = 1$ to get $\cos x$. Then $\tan x = \sin x / \cos x$.

9. (c) Use the addition formulae.

10. (a) Use $\cos^2 x = 1 - \sin^2 x$ and set $\sin x = y$.

10. (c) $\cos \theta° = \sin(90° - \theta°)$

11. Note $\sin(-x) = \sin x$ etc.

12. Use $a^4 - b^4 = (a^2 - b^2)(a^2 + b^2)$, $\sin^2 x + \cos^2 x = 1$, and the double angle formulae.

13. (a) Start with the cosine rule to get c then the sine rule to get another angle.

13. (b) Start with the sine rule. There are two possible solutions.

15. (a) Divide the pentagon into five equal triangles with a common vertex at the centre.

Coordinate Geometry

1. (b) Use (5.1).

1. (e) The shortest distance is along the line that is perpendicular to CD produced. Use the result of part (d).

4. Convert the two simultaneous conditions into one quadratic equation and look at its discriminant.

7. (a) Write $\beta^2 = pv^2 + q$ where p and q are constants you can determine from the given conditions.

8. (d) Set $\mu_1 d_1 = \mu_2 d_2$, where the subscripts $1, 2$ refer to lead and concrete respectively.

9. Compute $x^2 + y^2$.

11. (b) The slope of BC will be the same as the slope of AD which is known. So construct the equation for the line BC and seek its intersection with the equation for the side AB to get the coordinate B. For D you can use a similar procedure, or note that coordinate-wise $A + C = B + D$ (from the mid-point formula).

Plane Geometry

2. (a) The diameter AB passes through the centre O. Use the Inscribed Angle Theorem to deduce that $\angle ACB = 180°/2$.

2. (b) Let ACB be the triangle with $C = 90°$. By Pythagoras' theorem, the hypotenuse, c, will be the longest side in the triangle. Join vertices A and B to the centre O of the circle. Then by the

Inscribed angle theorem $\angle AOB = 2C = 180°$. That is, the line AB passes through the centre: It is the diameter.

3. Draw the radial lines to two non-adjacent vertices. Then use the Inscribed Angle Theorem twice to deduce a relation between the opposite interior angles and the angles at the centre.

4. Join the vertices with lines.

5. The hypotenuse is the longest side and it must be the diameter (See E2).

6. (a) AOD is an isosceles triangle and DAB is a right-angled triangle.

7. As an intermediate step you can imagine a point P on the circumference so that EP is tangent to the circle. Now apply the Tangent Secant theorem twice.

8. Let O be the centre of the circle of radius r, and let ABC denote the triangle with sides $a, b, c..$ Show that the area of triangle OBC is $ar/2$. Derive similar relations for triangles OAB and OAC. Sum them. (See also C4 in the Trigonometry chapter).

9. (a) Use the Intersecting Chords Theorem.

Differential Calculus

1. $V_x = dx/dt$.

4. Use the chain rule.

6. (b) Set $dV/dt > 0$.

13. Use (7.14).

Integral Calculus

2. (c) Sketch $\sin x$. Note the region where it is negative.

3. (f) Note $\sin 2x \cos 2x = (\sin 4x)/2$.

6. Integrate $dR/dt = 0.5$.

7. What is the maximum value of $\sin^2 x$?

9. (c) Solve $v(t) = 0$.

Appendix C

Answers to Exercises

Exponents and Logarithms

1. (a) -3.1×10^{-3}. (b) 1.23456×10^2. (c) 8.11376×10^{-1}.
   (d) -1.7017×10^{-1}

2. 1.3 s; 8.3 min; 4.2 yrs.

3. (a) 100. (b) 3.0 dB.

4. (a) $(\ln 2)/\lambda$. (b) 11460. (c) $\approx 24.500 \; yrs$. (d) Plutonium 239.

5. (a) mc^2. (b) c. (c) Yes. Photons.

6. (a) $1.2 \times 10^{-7} m$. (b) $3.6 \times 10^{-7} m$.

7. (a) $5^{7/2}$, 5^5, 7^5, 7^7. (b) $\log_2 \frac{1}{3}$, $\log_2 \sqrt{3}$, $\log_2 3$, $\log_2 9$.

8. (a) $4^{-\frac{3}{2}}$, $8^{1/3}$, $\left(\frac{1}{2}\right)^{-2}$. (b) $\left(2^2 \cdot 3^4\right)^{-\frac{1}{2}}$, $\left(2^{\frac{1}{2}}\right)^{-4}$, $2^2 \cdot 2^{-3}$, $\left(2^6\right)^{\frac{1}{3}}$

9. (a) $\dfrac{x^4}{3\sqrt{2}}$. (b) $\dfrac{1}{2} - \dfrac{25 \log_5 7}{2}$. (c) $\dfrac{7}{3} \log_3 5$

10. (a) $\dfrac{13\sqrt{3}}{2}$. (b) $\dfrac{29-11\sqrt{7}}{2}$. (c) $\dfrac{119\sqrt{2}}{12}+15$

11. (a) $-\frac{1}{2}$. (b) -5. (c) 5. (d) $\frac{51}{49}$. (e) 8 or $\frac{1}{8}$. (f) 5. (g) $3\ln 7$

12. (a) 1.1. (b) 67.9.

13. (a) $1+3x+\dfrac{15x^2}{4}$. (b) $1+18x^2+144x^4$. (c) $\dfrac{1}{x^8}+\dfrac{8}{x^6}+\dfrac{28}{x^4}$.
    (d) $64+320x+688x^2$. (e) $\dfrac{1}{x^5}+\dfrac{9}{x^4}+\dfrac{33}{x^3}$

14. (a) one. (b) $x\approx 0.81$

15. (a) two. (b) $x\approx 0.17$

Polynomials and Rational Functions

1. (a) $(1+\sqrt{5})/2$

2. (a) $t=2$. (b) $h_{max}=45/4$, $t=1/2$. (c) $(1+\sqrt{5})/2$.
   (d): (a) $3+\sqrt{21}$. (b) $h_{max}=35/2$, $t=3$. (c) $3+\sqrt{15}$

3. (a) $1-\dfrac{\sqrt{3}}{2}$. (b) $\dfrac{11-3\sqrt{13}}{2}$. (c) $\dfrac{-1\pm\sqrt{5}}{2}$.

4. (a) $\dfrac{-5-\sqrt{33}}{2}\le x\le\dfrac{\sqrt{33}-5}{2}$.
   (b) $4-\sqrt{13}\le x\le 4+\sqrt{13}$.
   (c) $x\ge 2(2+\sqrt{3})$ or $x\le 2(2-\sqrt{3})$.

5. (a) $7/2$, $1/2$, 7. (b) $\dfrac{7\pm\sqrt{41}}{4}$.

6. (a) $b=-10$, $c=25$. (b) $b=-1$, $c=-2$.

7. (a) $0\le y\le 25$. (b) $0\le y\le 9/4$.

8. (a) $0<b<8$. (b) $b=-1$. (c) $b<0$. (d) $b>8$.

9. (a) $1 \pm \sqrt{3}$. (b) $3(x^2 - 2x - 2)$. (c) $a = -1/4$.

10. (a) $f(x) = (x - 3)(x - 1)(x + 1)$. (c) $12c + 37 \geq 0$.

11. (a) $\dfrac{3}{(x + 2)} - \dfrac{1}{(x + 1)}$. (b) $\dfrac{-17}{(x + 2)} + \dfrac{4}{(x + 1)} + 5$.

    (c) $5x + \dfrac{4}{(x + 1)} - 3$.

Simultaneous Equations, Inequalities and Matrices

1. $A = 7$, $B = 5$.

2. $x < 25$.

3. (a) $5 \pm 2\sqrt{5}$.

4. $0 < x < 0.253572$, $y = 18 - x$.

5. Solve the trajectory equation simultaneously with the equation for the line.

6. (a) -2; $\frac{1}{2}\begin{pmatrix} -4 & 2 \\ 3 & -1 \end{pmatrix}$. (b) 0; no inverse.

   (c) 4; $\frac{1}{4}\begin{pmatrix} -2 & -1 \\ 2 & -1 \end{pmatrix}$. (d) 2; $\frac{1}{2}\begin{pmatrix} 1 & -3 \\ 2 & -4 \end{pmatrix}$.

7. (a) $x < \dfrac{1}{2}$. (b) $\dfrac{1}{2} < x < \dfrac{3}{2}$. (c) $x < \dfrac{3}{2}$.

8. (a) $x = -11/21$, $y = -17/21$.

   (b) $x = \dfrac{5 \mp \sqrt{145}}{12}$, $y = \dfrac{37 \mp 5\sqrt{145}}{24}$.

   (c) $x = \dfrac{713 + 75\sqrt{89}}{32}$, $y = \dfrac{133 + 15\sqrt{89}}{16}$.

   (d) $x = 1/9$, $y = -2/3$.

Trigonometry

1. (a) $21.8m$.   (b) $19.96°$.   (c) 9.75.   (d) The tan function is not linear.

2. (b) $\sin^{-1}\left(\frac{n_2}{n_1}\right)$.

3. (a) $y = x\tan\theta - \dfrac{gx^2 \sec^2\theta}{2U^2}$.   (b) $\dfrac{U^2 \sin 2\theta}{g}$

    (c) $R_{max} = \dfrac{U^2}{g}$; $\theta = \pi/4$.

4. (b) $1/160 > t > 1/480$.   (c) $0.0083 > t > 0.0074$.

5. (a) $\lambda = 2L/n$, $n = +$ve integer.

6. (c) $1/\sqrt{3}$, 1, $\sqrt{3}$.

7. (a) $\sin 35°$, $\cos 35°$, $\tan 55°$,   (b) $\csc 15° > \sec 15°$

8. (a) $0 \geq y \geq -1$.   (b) $2 \geq y \geq 1$.   (c) $1 \geq y \geq -1$.
    (d) $-1 \geq y \geq -5$.

9. (a) $\cos x = 0.4\sqrt{6}$; $\tan x = 1/(2\sqrt{6})$.
    (b) $\sin 2x = 0.16\sqrt{6}$; $\cos 2x = 0.92$; $\tan 2x = 4\sqrt{6}/23$.
    (c) $\sin(x + \pi/3) = 0.1 + 0.6\sqrt{2}$;
    $\cos(2x + \pi/4) = 0.46\sqrt{2} - 0.16\sqrt{3}$;
    $\tan(x + \pi/6) = \dfrac{\sqrt{3} + 2\sqrt{6}}{6\sqrt{2} - 1}$.

10. (a) 0.73.   (b) 1.84.   (c) $x° = 30°$.   (d) 0.14.

11. $f(0) = 5$, $f(-3) = 4$.

12. $\sin x = \sqrt{5/8}$; $\cos x = \sqrt{3/8}$; $\tan x = \sqrt{5/3}$.

13. (a) $c = 7.86$, $A = 38.8°$, $B = 61.2°$.
   (b) $B = 63.3°$ or $116.7°$, $C = 66.7°$ or $13.3°$, $c = 7.2$ or 1.8.
   (c) $a = 30.88$, $b = 34.9$, $C = 10°$.
   (d) $C = 10°$, $b = 11.3$, $c = 2.3$.

14. (*a*) 17.2. (*b*) 19.3 or 4.83. (*c*) 93.6. (*d*) 9.8.

15. (a) 2.38. (b) 3.63

16. 1.04.

17. (a) $330°$. (b) $150°$. (c) $270°$. (d) $200°$.

18. R^2.

Coordinate Geometry

1. (a) $(7/2,\ 6)$. (b) $y = 2x$. (c) $y = 13/2 - x/2$. (d) $1/2$.
   (e) $1/\sqrt{5}$.

2. (a) $y = -2x/3 + 13/3$. (b) $3x/2$.

3. (a) $P = (1/2, 0)$, $Q = (0, 1/3)$. (b) $R = (1/4,\ 1/6)$.
   (c) $y = 3x/2 - 5/24$. (d) Both $1/24$.

4. (a) $-12/5 < m < 0$. (b) $m = 0$ or $m = -12/5$.
   (c) $m > 0$ or $m < -12/5$.

5. T^2 vs R^3 or $\log T$ vs $\log R$.

6. (b) $T \propto a^{3/2}$ approximately.

7. (a) $\beta = \sqrt{1 - v^2/c^2}$. (b) $v = c$.

8. (b) Lead. (c) Lead: $\mu \approx 2.6 \; mm^{-1}$; Concrete: $\mu \approx 0.046 \; mm^{-1}$.
   (d) About 56 mm

9. $C = (0,0), \; R = A.$

10. (a) $(x-3)^2 + (y-7)^2 = 3^2$. (b) $(3,4)$.
    (c) $(x-3)^2 + (y+7)^2 = 3^2$. (d) $(x+3)^2 + (y-7)^2 = 3^2$.

11. (a) $(8/13, -1/13)$.
    (b) $B = (12/13, 5/13), \; D = (35/13, -19/13)$.
    (c) $18/13$.

12. (a) $a = 1, \; b = 3/2$. (b) $a = 6, \; b = -1$. (c) $a = 7, \; b = 6$.
    (d) $a = -3, \; b = -2$ or $a = 3, \; b = 4$. (e) $a = \dfrac{15 \pm \sqrt{281}}{2}$.

Plane Geometry

1. See the worked examples and our website.

2. Use the inscribed angle theorem

3. See worked example 2.

4. $180°$.

5. (a) 10. (b) $5\sqrt{3}$.

6. (a) $65°$. (b) $25°$. (c) $50°$. (d) $90°$.

7. $104/11$.

8. (a) $2\sqrt{14}$. (b) $2\sqrt{14}/7$.

9. (a) $ED = 7/3, \; DC = 3.33$. (b) $43.69°$. (c) $36.31°$.

Differential Calculus

1. $V_x = U \cos\theta$, $V_y = U \sin\theta - gt$.

2. (Is it time to practise deep breathing?)

3. (a) $6x - 6x^{-4}$. (b) $2/(2x+1)$. (c) $6\cos 3x$. (d) $-3\sin 2x$.
   (e) $3e^{2x}(1+2x)$. (f) $13/(x+5)^2$.

4. (a) $-\infty$. (b) 4. (c) 12. (d) 0. (e) 6. (f) 26/25.

5. (a) $6x + 9x^2 - 2/x^2$.
   (b) $\dfrac{3}{2\sqrt{2+3x}}$.
   (c) $\dfrac{-3(5x^3 + 4x^2 - 5)}{2(x^3+5)^2\sqrt{3x+2}}$.
   (d) $-x^2 \sin x + 2x \cos x + 2\cos 2x + \tan 5x + 5x \sec^2 5x$.
   (e) $x(x+2)e^x + 1/(x+1)$.
   (f) $\dfrac{\cos(\ln x)}{x}$.

6. (a) a: growth term. (b) $V > \left(\dfrac{b}{a}\right)^{1/(p-q)}$.

7. (a) $y' = 2ax + b$, $y'' = 2a$.
   (b) Turning point at $x = -b/(2a)$ is a local maximum if $a < 0$, a
   local minimum if $a > 0$.

8. (a) l.max at $x = -1 - 1/\sqrt{2}$, l.min at $-1 + 1/\sqrt{2}$.
   (b) g.min at $x = 0$. (c) g.min at $x = -1$, l.max at $x = 1$.
   (d) l.min at $x = 0$, inflexion pt at $x = \sqrt{\pi}$.
   (e) g.min at $x = -1/2$.

9. (a) $T : y = 4 - x$; $N : y = x + 2$. (b) $T : y = 7x - 2$; $N :$
   $7y + x - 36 = 0$.

10. (a) 1st curve: $-\infty \le x < 3/2$; no valid region for 2nd curve.

    (b) 1st curve: all x; 2nd curve, $x > -1/3$.

11. (a) $x = 0$, B. (b) $\dot{x} < 0$, population declines.

    (d) Stable equilibrium value of population.

12. (a) $2\pi k$. (b) $8\pi Rk$. (c) $4\pi R^2 k$.

13. (a) 0.628 cm. (b) 25.1 cm^2. (c) 125.6 cm^3.

14. (Time to take a stretch break?)

Integral Calculus

1. (a) 1. (b) 1. (c) $\ln 2$. (d) 1. (e) 2/3. (f) 1/3.

2. (a) 0. (b) 4.

   (c) In part (a) half the integral contributes a negative amount.

3. (a) $x^3 + 3x^4/4 - 2\ln x + C$. (b) $\dfrac{2(3x+2)^{3/2}}{9} + C$.

   (c) $-\cos 2x - 3\tan x + C$. (d) $\dfrac{2\ln(3x+5)}{3} + C$.

   (e) $(2/3)e^{3x} - (1+x)^4/4 + C$. (f) $-(7/8)\cos 4x + C$.

   (g) $-(2/3)\sqrt{2+3x} - e^{-2x}/2 + C$.

4. (a) 19.64. (b) 1.86. (c) 0.92. (d) 4. (e) 16.47. (f) 1.43. (g) 0.

5. $y = x - \ln x + 1$.

6. 12.5 cm

7. (Smile as you see the Hints appendix)

8. 142/81.

9. (a) $u - gt$. (b) $ut - gt^2/2$. (c) $t = u/g$. (d) $u^2/2g$. (e) $2u/g$.

   (f) $-u$.

   (g) Describes vertical particle motion in a constant gravitational field.

10. (a) -5. (b) Positive for some part of the range.

    (c) Negative for some part of the range.

π *goes on and on and on ...*
And e is just as cursed.
I wonder: Which is larger
When their digits are reversed?
(Anonymous, from the Web)

Appendix D

Answers to Odd-Numbered Problems

Exponents and Logarithms

1. (a) $N = N_0\, 2^{t/2}$; 6.64.

3. (a) 160.   (b) 103.7 mins.

5. (a) $\log_2 \sqrt{3}$, $\log_2 3$, $\log_2 6$, $\log_2 9$.

5. (b) $\log_4 3 = \log_2 \sqrt{3}$, $2\log_9 3$, $\log_2 2\sqrt{3}$.

7. (a) -1.48 or 3.08.   (b) 0.07.   (c) 0.67.   (d) 0.22 or 1.42.   (e) -3.

7. (f) 10.65.   (g) 2/3.   (h) 8.

9. $x = -1, y = 1$.

11. (a) 0.   (b) $(35 - \sqrt{73})/6$.   (c) $(1 \pm \sqrt{21})/2$.   (d) -1.

13. $\sqrt{3} + 1$

15. $10,927.27$

17. (a) 1, (b) 1, (c) 70, (d) 64.

19. 0.59

Polynomials and Rational Functions

1. (a) $x = 2^{1/4}$, $y = 1/x$.   (b) 297 mm by 210 mm.

3. (a) $a = 8 + 6\sqrt{2}$, $b = (17 + 12\sqrt{2})/50$.   (b) $(-200 + 150\sqrt{2}, 100)$.

3. (c) $100(3\sqrt{2} - 4)$.   (d) $3.55 \le x \le 20.71$.

5. (a) $\left(1 \pm \frac{1}{\sqrt{2}},\, 3 \pm \frac{1}{\sqrt{2}}\right)$.   (b) $(-3/2,\ 11/4)$ or $(1, -1)$.

5. (c) $(x, y) = \left(\dfrac{-1 \mp \sqrt{5}}{2},\ \dfrac{5 \pm \sqrt{5}}{2}\right)$.

7. (a) $\dfrac{\sqrt{49 - 8p}}{2}$, (b) $\dfrac{49}{2p} - 2$, (c) $\dfrac{8p^2 - 392p + 2401}{16}$.

9. $a = \sqrt{\dfrac{9 \pm 4\sqrt{2}}{7}} \approx 0.69;\ 1.45$.

11. (a) $V = 4\pi R^3/3$.   (b) $h = 1$. (c) $0 \le h \le 2R \Rightarrow h = 1$.

13. (a) $a = -1$, $b = -2$, $c = 2$.   (b) $Q = x$. (c) 1.

15. $p = \dfrac{15 \pm \sqrt{105}}{6}$.

17. (a) $-\dfrac{2x}{x^2 + 1} + x + \dfrac{1}{x}$.   (b) $\dfrac{-8}{x + 1} + \dfrac{4}{(x + 1)^2} + 5$.

17. (c) $\dfrac{17}{x + 2} - \dfrac{12}{x + 1} + \dfrac{4}{(x + 1)^2}$.

17. (d) $\dfrac{8x - 6}{5(x^2 + 1)} + \dfrac{17}{5(x + 2)}$.

19. 3/4.

Simultaneous Equations and Inequalities

1. $N = 52$; $(x = 30,\ y = 22)$

3. $x = 1, y = 5$.

5. (a) $\left(\dfrac{7 - \sqrt{61}}{6},\ \dfrac{\sqrt{61} - 1}{6} \right)$.   (b) $(-\sqrt{5},\ -1.)$

5. (c) $x = \dfrac{11 \pm \sqrt{105}}{2},\ y = \dfrac{11 \mp \sqrt{105}}{2}$

5. (d) $(0, 3)$ or $(1, 2)$.

7. (a) $2 < x < \dfrac{3 + \sqrt{41}}{4}$.   (b) $x > \dfrac{1 + \sqrt{21}}{2}$ or $x < \dfrac{1 - \sqrt{21}}{2}$.

7. (c) $x < 3/2$.

9. $x = 0.517638,\ y = 1.93185$.

11. (b) Cost per glass :

	Cool	Wow	Zing
Supplier A	1.4	1.1	1.1
Supplier B	1.2	1.0	1.1

(c) Apex's profit is $ 40. Bravo's profit is $ 43.5.

Trigonometry

1. 1125.4.

3. $A = 25$, $B = 5$, $k = 2\pi/50$, $C = \pi/2$

5. (a) $3/\sqrt{58}$.   (b) $\sqrt{\dfrac{1}{2} + \dfrac{7}{2\sqrt{58}}} \approx 0.98$.

7. (a) $-8/10$.   (b) $\dfrac{1}{10}\sqrt{50 + 5\sqrt{10}}$.

9. (a) -1.80. (b) 1.93.

11. $\pm R$.

13. (a) $1 \pm \cos A$.

15. $x = 0.63$, 1.88, 1.90.

17. (a) $-1/2$.   (b) $a = 3.76$, $b = 2.15$.

19. (a) $\pm\sqrt{58}$.   (b) $\pm\sqrt{5}$.

21. (a) 1.1. (b) 0.3. (c) 1.7.

23. (a) $0.5 \leq H \leq 6.5$.   (b) $t = (16/\pi) - 4$.

25. (a) $B = 17.9°$.   (b) $a = 10$.   (c) 8.5.

27. (a) To vertex $= 1/\sqrt{3}$; to side $= 1/(2\sqrt{3})$

27. (b) Height of pyramid $= \sqrt{2/3}$, angle $= 70.53°$

29. (a) 4.45 m. (b) 236.6 m.

31. $3\sqrt{3}/16$.

33. (a) 187.5 m. (b) 515.1 m.

35. (a) 2.92 m. (b) $240.3°$.

Coordinate Geometry

1. $y = 2e^{x^2/\sqrt{3}}$.

3. (a) $y = 5 - \sqrt{4 - z^2}$.

5. (a) $x = (-1 \pm \sqrt{13})/2$, $y = (2 \mp \sqrt{13})/3$.   (b) $13/3$.   (c) $\sqrt{13}/6$.

5. (d) $\sqrt{13}/13$.

7. (a) $B = (2, -2)$, $C = (-1, -5)$, $D = (-4, -2)$.   (b) 18.

9. (a) $R = \sqrt{2}$.   (b) $T_1 = (3, 2)$.

11. (a) $y = 2x + 3\sqrt{5}$.   (b) $\left(\dfrac{5 - 6\sqrt{5}}{5}, \dfrac{10 + 3\sqrt{5}}{5} \right)$.

11. (c) $y = (5 - x)/2$.   (d) $3\sqrt{5}/2$.   (e) $R = \sqrt{14}$. Center at P.

Plane Geometry

1. $(N - 2)\pi$.

3. $BC = 4.61$, $AD = 3.35$.

5. $36.27°$.

7. (a) $ED = 7/3$, $DC = 3.67$.   (b) $96.2°$.   (c) $126°$.

9. 22.69.

Differential Calculus

1. (a) $y' = 3ax^2 + 2bx + c$; $y'' = 6ax + 2b$. (b) $b^2 < 3ac$.

1. (c) $b^2 > 3ac$.   (d) $b^2 = 3ac$, $x = -b/3a$.

3. (a) $(2 + \cos 2x)\tan^2 x$. (b) $(2x^2 + 1)e^{x^2}$.

3. (c) $\dfrac{(x^2 + 3x + 1)\cos x - (2x + 3)\sin x}{(x^2 + 3x + 1)^2}$.

3. (d) $\dfrac{2x \cos \ln(1 + x^2)}{(1 + x^2)}$.   (e) $-2x \tan x^2$.

3. (f) $\dfrac{2(\ln 3x)(1 - \ln 3x)}{x^3}$.

5. (a) 0.015 cm.

7. (a) 22. (b) -1. (c) 0.54. (d) 17.52. (e) 18.78.   (f) $-1/4$.

9. (a) $t = \pi/12$. (b) -18.5. (c) $t = 0.124$. (d) -255. (e) $t = 0.248$.

11. (a) Square of side $L/4$.

11. (b) A rectangle with length but infinitesimal width.

13. (a) $\dfrac{2A \tan \theta}{(1 + \tan \theta)^2}$.   (b) $A/2$.

Integral Calculus

1. (a) $\pi/4$. (b) 0.   (c) 1.   (d) 2.

3. (a) $3\ln(x + 2) - \ln(x + 1) + C$.

3. (b) $5x + 4\ln(x + 1) - 17\ln(x + 2) + C$.

3. (c) $(5/2)(x+1)^2 - 8(x+1) + 4\ln(x+1) + C$.

3. (d) $5x - 4/(x+1) - 8\ln(x+1) + 5 + C$.

3. (e) $x^2/2 - \ln(1+x^2) + \ln(x) + C$.

5. (a) $4\sqrt{3}$. (b) $4/3$. (c) $4\sqrt{3} - 4/3$. (d) $(4\sqrt{2})/3$. (e) $4/3$.

7. (a) $e - 1 \approx 1.72$.   (b) $7/3$. (c) 9.38.

9. (a) $y = x + x^2/2 - 4\ln(1+x) - 1/2 + 4\ln 2$.

9. (b) $y = -(7/8)\cos 4x + C$, with $C = 1/8$ or $15/8$.

11. (a) $x(t) = 2 + 2t + 3t^2/2 - t^3/3$. (b) $t_1 = 5.58$. (c) $t_2 = 3.56$.

11. (e) 11.09.

13. (a) -2 m. (b) 6 m.

13. (c) From $t = \pi/2$ to $3\pi/2$ the particle travels in the $-x$ direction.

15. $14/3$

17. (b) 1.44.

19. 1.

21. $(\ln 2)/2 \approx 0.35$.

Appendix E

Applications Index

K

L

M

N

O

P

Q

R

S

T

V

Z

Appendix F

References

The quotations in this book by the first five notables below are from Wikiquote ($http://en.wikiquote.org/wiki/Main\_Page$) while the rest are taken from the MacTutor History of Mathematics archive of Dr John O'Connor and Prof. Edmund Robertson ($http://www-history.mcs.st-and.ac.uk/index.html$). The attribution of primary sources for the specific quotations follows.

Galileo: in The Assayer (1623), as translated by Thomas Salusbury (1661).
Einstein: unsourced variant of quote attributed to Einstein.
Euler: as quoted in In Mathematical Circles (1969) by H. Eves.
Descartes: Le Discours de la Mthode.
Newton: Letter to Robert Hooke.
Brahmagupta: Quoted in A History of Mathematics by F. Cajori.
Fermat: In the margin of his copy of Diophantus' Arithmetica.
Archimedes: Vitruvius, De Architectura ix, 215.
Plato: Supposedly inscribed above the doorway of his Academy.
Leibniz (no citation).
Al-Khwarizmi (no citation).
Sophie Germain: Letter to Gauss (1807).

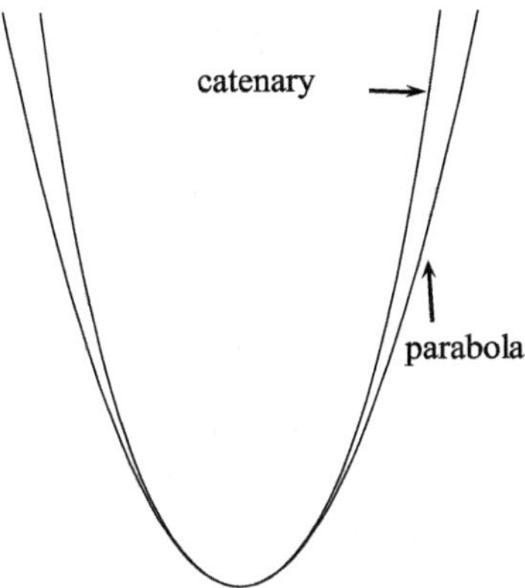

What shape would a suspension cable take?

The Authors

Adeline Ng

is a mathematics lover and explorer. She has used her vast teaching
experience, and personally tailored curricula, to enable students
achieve significant improvement in their mathematics competency
and reach their maximum potential. She also provides consultancy
and training for educational institutes.

Dr. Rajesh R. Parwani

is a theoretical physicist and mathematics enthusiast who specialises
in quantum theory and cosmology. He has taught science and
quantitative reasoning modules at a university, and has mentored
research students from secondary to graduate school. Currently, he
provides research and education consultancy.

Contact email: enquiry@simplicitysg.net

Print Edition SRI-2013-3A
(Errata Corrected Dec 2015)

www.ingramcontent.com/pod-product-compliance
Lightning Source LLC
Chambersburg PA
CBHW050509160726
48003CB00001B/230